(DIS)PLACING EMPIRE

T0230566

For Jenny and Caroline

(Dis)Placing Empire
Renegotiating British Colonial Geographies

Edited by

LINDSAY J. PROUDFOOT
Queen's University, Belfast

MICHAEL M. ROCHE
Massey University, New Zealand

Routledge
Taylor & Francis Group

LONDON AND NEW YORK

First published 2005 by Ashgate Publishing

Published 2017 by Routledge
2 Park Square, Milton Park, Abingdon, Oxfordshire OX14 4RN
711 Third Avenue, New York, NY 10017, USA

First issued in paperback 2017

Routledge is an imprint of the Taylor & Francis Group, an informa business

British Library Cataloguing in Publication Data
(Dis)placing empire : renegotiating British colonial
  geographies. - (Heritage, culture and identity)
  1. Imperialism 2. Postcolonialism 3. Historical geography
  4. Great Britain - Colonies - Social conditions 5. Great
  Britain - Colonies - History 6. Australia - Emigration and
  immigration - Social aspects 7. Australia - Social
  conditions
  I. Proudfoot, L. J. (Lindsay J.) II. Roche, M. M.
  325.3'41

Library of Congress Cataloging-in-Publication Data
(Dis)placing empire : renegotiating British colonial geographies / edited by Lindsay
J. Proudfoot and Michael M. Roche.
      p. cm. -- (Heritage, culture and identity)
  Includes bibliographical references and index.
  ISBN 0-7546-4213-5
  1. Great Britain--Colonies--History--19th century. 2. Great Britain--Colonies--
History--20th century. 3. Postcolonialism--English-speaking countries. 4.
Decolonization--English-speaking countries. 5. Postcolonialism--Commonwealth
Countries. 6 Decolonization--Commonwealth countries. 7. Ireland--History--19th
century. I. Title: (Dis)placing empire. II. Proudfoot, L. J. (Lindsay J.) III. Roche,
M. M. IV. Series.

  DA16. D57 2005
  909'.0971241081--dc22

                                                                        2005003566

ISBN 13: 978-1-138-27468-6 (pbk)
ISBN 13: 978-0-7546-4213-8 (hbk)

Contents

PART III: 'DISPLACEMENT'

List of Figures

List of Tables

List of Plates

List of Contributors

Paddy Duffy is Associate Professor of Geography in the National University of Ireland, Maynooth where he is Director of postgraduate studies. Author of *Landscapes of south Ulster: a parish atlas of the diocese of Clogher* (1993). Editor of *To and from Ireland: planned migration schemes, c1600-2000* (2004), and co-editor, *Gaelic Ireland c1250-c1650: land, lordship and settlement* (2001, 2004).

Di Hall has held research fellowships at Queen's University, Belfast and the University of Melbourne where she now holds an Australian Research Council Post-doctoral Fellowship in the History Department.

Philip Howell is a University Lecturer in Geography at Cambridge and a Fellow of Emmanuel College. His recent research focuses on the regulation of prostitution in 19th-century Britain and its empire.

M. Satish Kumar teaches in the School of Geography, Queen's University, Belfast. He was the recipient of the prestigious Commonwealth Fellowship to Cambridge and was recently awarded the Bhoovigyan (Earth Scientist) National Leadership Award for contributions to Population, Environment and Development Studies. He has jointly edited *Globalisation and North-East India: Some Developmental Issues* (in Press) and is the author of the forthcoming book, *Urbanising the Developing World: From a Colonial to Post-Colonial Context*. With others he is also completing a volume entitled *Colonial and Postcolonial Geographies of India.*

Joy McCann is a historian who works in the field of Australian cultural history and heritage. She has a deep interest in rural landscapes and community-based environmental and social histories. Her doctoral research at the Australian National University explores intersections between farming landscapes, rural decline and social memory in Australia's wheatlands.

J.M. Powell, MA (Liverpool), PhD, DLitt (Monash), Fellow of the Academy of Social Sciences in Australia and Corresponding Fellow of the British Academy, former president of the Institute of Australian Geographers and Editor of *Australian Geographical Studies*, is currently Emeritus Professor of Historical Geography at Monash University. He has served on numerous international editorial boards including the *Journal of Historical Geography*, and has taught and researched in the United Kingdom, Western Europe, North America, Australia and New Zealand. His most recent work focuses on conservation and resource appraisal in the British Empire/Commonwealth between 1900 and 1950, geographers and the geographical imagination in modern Australia, and national and regional environment-identity convergences.

Dr Lindsay Proudfoot is a Reader in Geography at Queen's University, Belfast, specializing in the Historical Geographies of Post-Plantation Ireland and the Irish in the British Empire. His major publications include: *An Historical Geography of Ireland* (1993, with B. Graham); *Urban Patronage and Social Authority. The Management of the Duke of Devonshire's Towns in Ireland 1760–1890* (1995); *Property Ownership and Urban and Village Improvement in Provincial Ireland, ca. 1700–1845* (1997); and *Provincial Towns in Early Modern England and Ireland. Change Convergence and Divergence* (2002, with P. Borsay). He directed the Leverhulme Trust funded project, 'Memory, place and symbol: Irish identities and landscape in colonial Australia', between 2001–2004.

Michael Roche is Professor of Geography in the School of People Environment and Planning at Massey University in Palmerston North, New Zealand. His recent research has focused on society and environment in New Zealand in the 1920s and 1930s. Previously he has written on forest conservation and the timber industry, the growth of the frozen meat industry and the emergence of export fruit growing in the Southern Hemisphere. He has contributed to the *Historical Atlas of New Zealand* (1997), various volumes of the *Dictionary of New Zealand Biography* as well as a chapter to *Environmental Histories of New Zealand* (2002).

Yvonne Whelan lectures in cultural geography at the Academy for Irish Cultural Heritages at the University of Ulster. Her research interests focus on the relationship between landscape, memory and the politics of identity. Her book, *Reinventing modern Dublin: landscape, power and the politics of identity,* is forthcoming.

Preface

The idea for this volume originated at the International Conference of Historical Geographers meeting in Quebec in 2001. The Conference has existed in various guises and has been held at various locations for 30 or so years, and has proved to be a valuable focal point for the exchange of ideas between historical geographers from different parts of the world. Indeed the lack of an over-arching formally structured organization underpinning the Conference has had the advantage of ensuring that these meetings have embraced multiple perspectives which have reflected their peripatetic locations. However, one consequence has been the absence of a conference proceedings. By appearing individually as journal articles, contributions to sometimes substantial debates around common themes have lost their sense of shared perspective and coherence. By bringing together a number of contributions on the spatiality and materiality of colonialist experiences of Empire in book form, the editors have sought to remedy this deficiency by providing a substantial commentary on one of the major themes of the Quebec meeting. The essays include work by a number of younger scholars as well as by some more established figures in the field, and have been supplemented by additional contributions from authors who were not at the conference but who offer contingent post-colonial perspectives. The range of material is considerable. The time/place moments covered by the authors vary from pre-Raj India to 18th-century Ireland, 19th-century Hong Kong, 1920s New Zealand and modern Australia. Some omissions in a collection of this sort are inevitable. In this instance, neither Canada nor South Africa figure in the contributions. Nevertheless, the diversity of material covered and the range of perspectives offered, aptly reflect the heterogeneity of the 'white' experience of Empire that is the book's fundamental *locus*.

Acknowledgements

The editors wish to acknowledge the assistance of Olive Harris (formatting and layout) and Maura Pringle (cartography). Jenny Proudfoot willingly agreed to proofread the text and Dr Ian Fuller helped scan some of the images to the publishers' specifications. The permission of the National Library of Ireland, the National Library of New Zealand and of the Alexander Turnbull Library (New Zealand) to reproduce images from their collections is both acknowledged and appreciated.

Lindsay Proudfoot and Dianne Hall wish to acknowledge the generous funding of the Leverhulme Trust in supporting the project 'Memory, Place and Symbol: Irish Identities and Landscape in Colonial Australia', on which chapters 4 and 5 are based.

Joy McCann acknowledges permission from the Australian Bureau of Agriculture and Resource Economics to reproduce Figure 3.1, and from the New South Wales Environmental Protection Agency to reproduce Figure 3.2.

Chapter 1

Introduction: Place, Network, and the Geographies of Empire

Lindsay Proudfoot and Michael Roche

'When, and where, was Empire?' asks Philip Howell in his contribution to this volume. To which we may add, what was Empire? Diverse in practice, fractured and ambiguous in meaning, local yet transnational, European imperialism, particularly as formulated in the British Empire, continues to inflect the lives and understanding of tens of millions of people through its post-colonial legacies in the contemporary world. Geographers are no exception. Among the various 'turns' that have characterised the production of Human Geography in the Anglophone academy since the 1980s, the 'postcolonial turn', with its attendant emphasis on exposing Geography's disciplinary complicity in the practice of Empire, de-coupling local (post)colonial geographies from Westernising master-narratives, and recovering the spaces of the 'colonised', constitutes the latest epistemological shift to have been credited with potentially revolutionary significance within the discipline (Philo, 2000; Yeoh, 2000; Ryan, 2004). Yet to describe the postcolonial turn in Geography in this way is to risk appearing unduly dismissive. These geographical studies reflect and form part of a much broader critical intellectual 'moment' in a variety of disciplines, notably anthropology and literary and cultural studies. Although diverse in its language, pre-occupations and self-identification, this 'moment' shares a general politicized textual concern with the historical effects of imperialism and colonialism on people and places, and with the 'reproduction and transformation of colonial relations, representations and practices in the present' (Gregory, 2000, p. 612; see also Loomba, 1998; McLeod, 2000; Young, 2001).

As Clayton (2004, p. 250) points out, central to these concerns is the idea that culture, conceived of in terms of 'language and knowledge, texts and discourses, images and representations, and the iconography of power', has been as important in the projection of Western power as 'the political economy of imperial expansion and colonial incursion'. Implicit within this are Foucauldian notions concerning knowledge-power relationships; hence Western imperialism and its attendant colonialisms are represented as acts of epistemological aggression (Said, 1978; Lester, 2000; Edgar and Sedgwick, 2002). In asserting the primacy of the Western intellectual tradition and its forms of knowledge, imperialism embodied its own justification through its denial of the validity of alternative indigenous cultures and knowledges. Thus the language of imperialism, grounded in an Enlightenment narrative of modernity which privileged the European 'West' as the sovereign centre of this particular world view, became both the arbiter of truth and the rationale for dominance (Said, 1978; Kennedy, 1996; Walia, 2001). These Eurocentric epistemologies are held to have survived the political decolonization of the former spaces of Empire and to have continued to subvert the construction and representation of non-Western identities. Accordingly, a primary concern of postcolonial scholarship has been to 'decentre' such Westernising intellectual constructions both within and beyond the Western academy (Guha, 1996; Pierterse and Parekh, 1995). Consequently, Crush (1994) and others (McEwan, 1998; Sidaway, 2000) have argued that Geographers should be concerned to 'decolonize' their subject, to make explicit the unspoken, racialized, patriarchal Eurocentric assumptions that have asserted the primacy of Western forms of Geographical understanding, and which continue to impede attempts to articulate the spaces of indigenous identity (see also Ryan, 2004).

Yet despite this alertness to the continuing Western 'intellectual colonization' of the former spaces of Empire, the question that still haunts postcolonial studies and was most famously posed by Gayatri Spivak (1988) is, 'can the subaltern speak?' In 'giving voice' to subaltern indigenous identities, are postcolonial scholars merely denying the authenticity of their 'silences', and in any event simply re-construing their 'Otherness' in terms which reaffirm the primacy of the Eurocentric intellectual perspectives that they are ostensibly seeking to subvert? At the heart of this issue is the question of whether it is possible for postcolonial scholarship ever to be written from 'outside' its own disciplinary history and cultural tradition, and thus escape the totalizing Western constructions of modernity and colonial difference which imbue these (Chatterjee, 1999; Prakash, 1996, 1999). Even though we may doubt whether such intellectual decolonization is ever fully attainable, and that national and cultural identities can ever be identified in the former spaces of Empire which are truly untainted by colonialism, this post-modern concern with authorial positionality has been identified as one of the defining characteristics of postcolonial scholarship. Recent commentaries on imperial and postcolonial geographies have argued that it is no longer possible to write from the erstwhile imperial 'metropolitan centre' in an un-reflexive way which ignores the constitutive assumptions of dichotomy and binarism inherent in this positionality (Ryan, 2004). We are adjured to be alert to the constructive nature of difference, even though we may be unable fully to disentangle ourselves from its consequences.

The contributions brought together in this book reflect this concern with authorial situatedness, but also mirror the eclecticism that has characterised recent postcolonial approaches in Geography (Clayton, 2004). Most of the contributions derive from papers given at the Eleventh International Conference of Historical Geographers (Quebec, August 2001), where discussions on imperialism and colonialism highlighted the need to engage with the material practices of colonialism, and acknowledge the constructive importance of geographical ideas of place and space in accounting for Empire's discursive complexity (see also Harris, 2002; Nash, 2002). The book's primary purpose is thus to deconstruct totalizing accounts of Empire, whether they be framed in terms of a conventional historicist

Western binary, or in terms of the, arguably, equally blunt and dichotomous lens of postcolonial discourse analysis. In doing so, the contributors privilege, the inherent materiality and spatiality of global human experience, but in terms which recognise that these can reflect broader, subjective, realities. These issues are addressed using a variety of lenses and perspectives to explore the cultural meanings inherent in the material practices of colonialism exhibited in various situated 'sites of colonization' as well as in the discursive spaces of the 'second' British Empire, in Ireland, Asia and Australasia during the later nineteenth and twentieth centuries. Central to these formulations is the idea that the individual experience of Empire was mediated through the interaction between the local specificities of 'place' and the externalities inscribed in these wider imperial networks or circuits (Thomas, 1994; Lester, 2000; Bridge and Fedorowich, 2003). 'Places' are conceived of here as loosely-bounded material sites of memory, agency and identity, activity-nodes in the life paths of ethnically, socially and culturally diverse groups of individuals, who brought their own understandings and gave their own meanings to the symbolic material spaces they encountered there (Anderson and Gale, 1992; Daniels, 1992; Entrikin, 1994, 1996, 1997). As polyvocal and inchoate sites of meaning, places were central to the formation, reproduction and contestation of individual and collective identities, of 'selfhood' and, consequently, alterity, among both the colonizing and colonized peoples of Empire.

But as sites of identity and interaction, these colonial places were also partly constituted by their relationship with other places, and were connected to them by discursive flows of information, knowledge and belief, as well as by more material flows of capital, commodities and labour. In his analysis of the Cape Province in the early nineteenth century, Lester (2001) has argued that such flows articulated broader imperial discourses, among which he identifies settler capitalism, governmentality, and (evangelical) humanitarianism as the most profoundly influential in this, as in other parts of the British Empire. He suggests that each represented 'a particular ensemble of regulatory practices', which were sufficiently coherent to be considered to be a separate imperial project, and which were 'devised and prosecuted by different British interests' in different colonial spaces, and which accordingly, encountered various forms of indigenous resistance. (Lester, 2001, pp. 3-5). The important point, however, as the chapters by Kumar and Howell in this volume demonstrate, is that these discursive flows were neither solely centrifugal, and predicated on disseminating information and ideas from the imperial 'core' to the colonial periphery, nor centripetal, exclusively acting to exhibit social and other advances devised in the colonial 'laboratories of empire' at the core. Rather, not only were they reciprocal *between* the imperial metropolitan core and the colonial periphery, but they also linked different sites *within* the colonies in similarly constituted, spatially and temporally asymmetrical relationships of exchange (Bridge and Fedorowich, 2003). Nor were these discursive flows and spaces in any way essentialist or exclusive. The imperial metropolitan and colonial interests represented by these discourses were constantly reforming, and thus individuals might 'move between different discourses, appropriating their rhetoric, and thus…transgressing the boundaries of different imperial networks' (Lester, 2001, p. 5).

This view of Empire as a web of mutually constitutive discursive networks, helps to break down the binarism which lay at the heart of the historicist Eurocentric imperial/colonial world view. Other contributions to this volume further deconstruct this essentialist binary of core/periphery, self/other, Western modernity/aboriginality, through their emphasis on the heterogeneity of the 'white' experience of Empire, either in the settler-capitalist dominions, or through the exploration of contested sites of identity in what was arguably Britain's oldest and most ambiguous colony, Ireland. Emphasizing the time/place contingency of these colonial encounters and their frequently discursive character, they respond to Thomas's claim that:

> The dynamics of colonialism cannot be understood if it is assumed that some unitary representation is extended from the metropole and cast across passive spaces, unmediated by perceptions or encounters. Colonial projects are constructed, misconstructed, adapted and erected by actors whose subjectivities are fractured, half here, half there, sometimes disloyal, sometimes almost 'on the other side' of the people they patronize and dominate, and against the interests of some metropolitan office. Between the Scylla of mindlessly particular conventional colonial history...and the Charybdis of colonial discourse theory, which totalizes a hegemonic global ideology, neither much tainted by conditions of production nor transformed by colonial encounters or struggles, lies another (ethnographic) path, which presupposes the effect of larger objective ideologies, yet notes their adaption in practice, their moments of effective implementation as well as those of failure and wishful thinking (1994, pp. 60-61).

Like Kumar and Howell, these contributors position themselves very much as writing 'for and from the edge of Empire', but the 'colonial encounters' and 'fractured subjectivities' they explore are framed primarily in terms of 'colonialism's geographies' (Clayton, 2004, p. 459). In these 'colonies of settlement' (to which Ireland, spatially proximate to the metropolitan core of Empire yet culturally distant, bore recognisable similarities as well as significant differences), imperial/colonial discourses were shaped rather differently from those in the African and Asian 'colonies of exploitation' (Bridge and Fedorowich, 2003). Imperial hegemony and colonial subalterity remained at issue, but were as likely to be politicized in terms of settler ethnicity and cultural difference as racialized in terms of a 'white/non-white' Other (Trainor, 1994; Gascoigne, 2002; Lester, 2002; Patterson, 2002; Jones and Jones, 2003; Lowry, 2003). 'Silences' existed, and in their eloquence might bear witness to the displacement of indigenous peoples and to the erasure or debasement of their cultures, but they could also testify to discourses of alienation and marginality within the European settler stream (O'Farrell, 1988; Birch, 1996; Gelder and Jacobs, 1998; Proudfoot, 2003). The sheer number of European, primarily British and Irish, emigrant settlers also told in the complex landscape inscriptions formed and re-formed in environments that were frequently hostile and inevitably alien (Crosby, 1986; Griffiths and Robin, 1997; Harris, 1997; Seddon, 1997).

The chapters in this volume thus traverse an intellectual terrain that is inflected with the European imperial arena and Eurocentric knowledge as its prime referents (Clayton, 2004, p. 258). Although a number of contributors (Duffy, McCann,

Roche and Whelan) allude to indigeneity in various contexts, only Kumar and Howell are primarily concerned with the recovery of (imperially prescribed) indigenous spaces. In their common concern with the heterogeneous 'settler spaces' of Empire, the other authors offer a qualified response to Bridge and Fedorowich's recent call for the re-assertion of an unashamedly 'British' emphasis in the history of Britain's imperial encounter with the modern world:

> Blinded by national historiographies and mesmerized by the exotic colonial 'other' we have lost contact with what was always the heart of the imperial enterprise, the expansion of Britain and the peopling and building of the trans-oceanic British world. It is time we reacquainted ourselves with what was once considered both vitally important and self-evident (2003, p. 11).

While we may question the mesmeric qualities of the postcolonial concern with ethnic, cultural and racial difference in imperial and post-colonial contexts, and note once again Ireland's ambiguous position in this formulation, Bridge and Fedorowich are surely right to emphasize the centrality of the 'neo-Britains' and 'neo-British' experience in the construction of Empire, and hence to any contextualized understanding of the cultural consequences of its formal political demise (Belich, 1996, 2001). Yet to assert this is not to claim some essentialist hegemonic view of 'Britishness' which marginalizes the histories and identities of the colonized peoples of Empire. Rather, as Pocock (1975, 1999) argues and the essays in this collection affirm, it is to recognise both the cultural and political pluralism inherent in the histories of the archipelagic nations from which the 'neo-British' sprang, and their need to redefine their own sense of their shared 'British' history in the face of its continuing renegotiation in the erstwhile imperial heartland.

The Organization of the Book

The contributions to this book are grouped together in two sections. The first section, 'Dis-locations', contains the five chapters that have as their primary concern, explorations of the contested identities and meanings embedded in the material practices associated with particular sites of colonization in Ireland and Australia. Themes of collective and individual resistance loom large here, while various broader imperial discourses provide further context: land ownership and improvement (Duffy), environmentalism (McCann), religious observance (Proudfoot), gender and domesticity (Hall), and ritual and spectacle (Whelan). Patrick Duffy offers an account of the 'estate' system of land-ownership in post-Plantation Ireland, which reads this as a series of hegemonic colonial spaces of surveillance and subaltern resistance. He argues that these reflected the ambiguities inherent in Ireland's position within Empire, as both complicit partner in various projects of imperial governmentality, and subordinated, colonized, 'indigenized' client-nation in a recent constitutional union (Lloyd, 1999; Howe, 2000; Graham, 2001). Underpinned by a moral discourse of civilizing improvement which misread local strategies of subaltern resistance as evidence of pre-colonial aboriginal disorder, Ireland's estates with their demesnes and 'big houses' were sites of inscription for both elite colonial 'Selfhood' and the 'Other' of a continuing, competing discourse of indigenous identity.

Similar themes of land possession, identity inscription and indigenous dispossession underpin Joy McCann's exploration of the complex inter-weavings between landscape, history and memory in the liminal spaces of Australia's wheat belt. Situating her account in the Lachlan Valley of central-west New South Wales, she examines the contrasting and contested ecological and cultural mythologies inscribed in the same landscapes by local indigenous groups and 19th century 'pioneer' settlers and their descendents. Carrying her narrative through to the present day, McCann demonstrates the inchoate nature of the non-indigenous understandings of place in particular. She argues that these 'white settler landscape mythologies' have been constantly re-invented in order to legitimize both the colonizing process and the massive environmental damage and indigenous displacement that ensued from this.

This theme of the instability and hybridity of place-based settler identities is developed by Lindsay Proudfoot in the slightly earlier context of pioneering Scottish Presbyterian missionary activity in 19th century Victoria and New South Wales. He argues for the general importance of religious affiliation as a major driver of settler identities and, accordingly, as a source of cultural inscription in the 'white spaces' of Empire. He uses an exploration of the agency of one particular foundational figure in Australian Presbyterianism to demonstrate the intensely personal nature of these inscriptions, and the fractured ways in which they might transgress the 'public' and 'private' spheres. Similar transgressions between differently inscribed public and private spaces form the subject of Dianne Hall's discussion of the construction of Irish women's identities in the central-western Victorian goldfield township of Glenorchy, in the early 1860s. She uses newspaper accounts of a locally-notorious incident of domestic violence between the mistress of one of the township's hotels and her female servant, to explore the gendered construction of colonial domestic space and the ways this underpinned social legitimacy. Hall demonstrates how, in challenging the feminised hierarchy of domestic space, individual behaviour could be seen to subvert this legitimacy and threaten emplaced collective identities.

In the final contribution to this section, Yvonne Whelan situates the theme of place identity in a discussion of imperial spectacle as a legitimising discourse, taking as her example Queen Victoria's visit to Dublin in 1900. She acknowledges, with Duffy, the ambiguities of Ireland's role in Empire, and shares his concern with the acts of subaltern resistance embedded in the country's colonial geographies. Whelan shows how the Queen's visit generated its own fluid spaces of imperial urban theatre, and how these were read in contradictory ways by republican, nationalist and unionist textual communities, who were united only in their shared acknowledgement of the importance of the visit's political symbolism. In uncovering both the geographies of imperial loyalism and resistant nationalism exhibited during the Queen's visit, Whelan demonstrates the hybridity of ostensibly imperial urban geographies near the heart of Empire. In so doing, she reminds us, with David Thomas (1994), that the meanings of Empire were neither stable nor homogenous, but were continuously renegotiated at the Empire's core as at its periphery (Thompson, 2000).

The four chapters that comprise the second section, 'Translocations', focus on the broader discursive networks and spaces of Empire rather than the situated place identities of the previous contributions. Nevertheless, they engage with similar material themes. Thus J.M. Powell shares Joy McCann's concern with the Australian immigrant nation's encounter with the enigmas of its environment. Specifying this encounter in a national frame, he examines the ways in which these enigmas were imagined, understood, represented and exploited, and how these engagements in turn helped validate a multiple and contested sense of 'national' identity and belonging which enhanced the country's sense of distinctiveness from Britain. He narrates the growth of a far from uncontested environmental awareness that first invoked and then destabilised various national myths, for example that of the 'heroic' pioneer, but which nevertheless invited an imaginative appropriation of the new country by art and literature as well as other representations. This act of appropriation also involved the exorcism of the convict stain and the silencing of indigenous memory in the moral agendas of Australian progress and civilisation, each of which had inscribed their mark on the landscape.

Michael Roche's account of the political debates surrounding Discharged Soldier Settlement in New Zealand during and after World War I, shares a similar concern with territorial discourses of national identity. He explores the Soldier Settlement debates in terms of their embedded moral rhetoric of Empire and Nationhood, duty and sacrifice. Roche examines the ways in which this political rhetoric, grounded in a colonial imaginary which privileged land as a source of national identity, became inflected and dominated by various contradictory social, economic, racial and class discourses. He argues that while the plans for Soldier Settlement initially embodied notions of the 'National debt' owed to soldiers who had fulfilled New Zealand's obligations to Empire, latterly its spaces became much more ambiguous, eliding such imperial and nationalist ideals with less comfortable sentiments of racial, gender and class difference.

Roche's emphasis on land occupation as part of the moral narrative of national identity, and on the exclusion of 'aliens' from this, echoes the subtext of dispossession in Duffy's account of colonial landownership in Ireland and in McCann's exploration of representations of pioneer land occupation in New South Wales. It also invokes the same sense of racialized imperial space which underpins the two final chapters in the book: Kumar's account of prostitution and its regulation in the urban spaces of colonial India, and Howell's exploration of the discursive contexts of prostitution regulation as an expression of imperial governmentality with reference to 19th century Hong Kong. Both authors emphasize the bounded yet permeable racialized spaces of surveillance this regulation created in these different parts of the 'non-white' Empire. Thereafter, each author offers complimentary perspectives. Kumar considers the local geographies of regulated prostitution, unpacking the micro-geographies of officially sanctioned sexual provision in Indian military cantonments in particular. Additionally, he also traces the origins of Indian prostitution in the pre-colonial era, offering an account of its transmogrification from social necessity to commodified neo-criminality under the British Raj. Howell contrasts the official discourse of regulationism, which insisted that regulation was a local response to colonial conditions, with the contemporary discourse of repeal, which rejected such claims to 'local necessity'. Constructing instead a powerful, and ultimately prevalent, view of regulation and its associated contagious disease regimes as a morally bankrupt system of imperial control, this discourse of repeal ultimately facilitated the demise of these regulated geographies of promiscuity.

End Note

This collection of chapters offers a series of diverse perspectives on the meaning and practice of Empire, as negotiated at various times and in various places, both by first-generation British and Irish emigrants and their immediate descendents, and by the 'neo-Britons', who (re)populated the 'white dominions' in the early-mid 20th century. For millions of the latter, their experience of Empire brought them into contact primarily with people who were remarkably similar to, yet quite unlike, themselves. For such, colonial difference was as likely to be constructed in terms of environmental learning, and in the nuanced rendition of subtle social, ethnic and cultural dissimilarities within the settler community, as in any overtly racialized discourse of indigeneity. The sad and eloquent fact is that by the early 20th century, in places such as Canada and Australia, indigenous peoples had already been largely relegated to the marginal spaces of Western 'Modernity'. Elsewhere, of course, in Africa and Asia, such racialized interactions continued to be much more pervasive in imperialism's complex inscriptions and in the representation of their postcolonial aftermath. In Ireland, in a perhaps rare valid example of the exceptionalism that has characterised that country's historiography, geographical propinquity to the imperial metropolis, economic disadvantage, and prescriptive 'pre-modern' cultural authenticity, combined to create an unusually ambiguous transgressive colonial state.

Such was the diversity of the European experience of Empire and its encounter with the colonial "Other' which forms the main concern of this book. By its very nature, of course, the heterogeneity of this experience renders any claim to a meaningfully panoptic view, suspect. Nor do the authors claim such Olympian detachment. Writing from various 'places' within the post-colonial world, they offer an understanding of aspects of the meaning and practice of Empire which is framed in terms of their own subjectivities. The lens they provide brings certain subjects and issues into sharp focus, leaving others as contextual background. The view is theirs; the shared experience of the legacies of Empire, ours.

References

Anderson, K. and Gale, F. (1992), 'Introduction', in K. Anderson and F. Gale (eds), *Inventing Places. Studies in Cultural Geography*, Longman, Melbourne, pp. 1-12.

Belich, J. (1996), *Making Peoples. A History of the New Zealanders. From the Polynesian Settlement to the End of the Nineteenth Century*, Allen Lane, Auckland.

Belich, J. (2001), *Paradise Reforged: A History of the New Zealanders. From the 1880s to the Year 2000*, Allen Lane, Auckland.

Birch, T. (1996), '"A Land So Inviting and Still Without Inhabitants": Erasing Koori Culture from {Post-) Colonial Landscapes', in K. Darian-Smith, L. Gunner and S. Nuttall (eds), *Text, Theory, Space. Land, Literature and History in South Africa and Australia*, Routledge, London, pp. 173-90.

Bridge, C. and Fedorowich, K. (2003), 'Mapping the British World', *Journal of Imperial and Commonwealth History*, vol. 31, pp. 1-15.

Castle, G. (ed.) (2001), *Postcolonial Discourses. An Anthology*, Blackwell, Oxford.

Chatterjee, P. (1999), *The Partha Chatterjee Omnibus*, Oxford University Press, Oxford.

Clayton, D. (2004), 'Imperial Geographies', in J.S. Duncan, N.C. Johnson, and R.H. Schien (eds), *A Companion to Cultural Geography*, Blackwell, Oxford, pp. 449-468.

Crosby, A. (1986), *Ecological Imperialism. The Biological Expansion of Europe, 900-1900*, Cambridge University Press, Cambridge.

Crush, J. (1994), 'Post-colonialism, De-colonization, and Geography', in A. Godlewska and N. Smith (eds), *Geography and Empire*, Blackwell, Oxford, pp. 333-50.

Daniels, S. (1992), 'Place and the Geographical Imagination', *Geography*, vol. 77, pp. 310-22.

Duncan, J.S. and Lambert, D. (2004), 'Landscapes of Home', in J.S. Duncan, N.C. Johnson, and R.H. Schien (eds), *A Companion to Cultural Geography*, Blackwell, Oxford, pp. 382-403.

Edgar, A. and Sedgwick, P. (2002), *Cultural Theory. The Key Thinkers*, Routledge, London.

Entrikin, N. (1994), 'Place and Region', *Progress in Human Geography*, vol. 18, pp. 227-33.

Entrikin, N. (1996), 'Place and Region 2', *Progress in Human Geography*, vol. 20, pp. 215-21.

Entrikin, N. (1997), 'Place and Region 3', *Progress in Human Geography*, vol. 21, pp. 263-68.

Gascoigne, J. (2002), *The Enlightenment and the Origins of European Australia*, Cambridge University Press, Cambridge.

Gelder, K. and Jacobs, J. (1998), *Uncanny Australia. Sacredness and Identity in a Postcolonial Nation*, Melbourne University Press, Melbourne.

Graham, C. (2001), *Deconstructing Ireland. Identity, Theory, Culture*, Edinburgh University Press, Edinburgh.

Gregory, D. (2000), 'Postcolonialism', in R.J. Johnston *et al.* (eds), *Dictionary of Human Geography*, Blackwell, Oxford, pp. 612-15.

Griffiths, T. and Robin, L. (1997), *Ecology and Empire*, Keele University Press, Keele.

Harris, C. (2002), *Making Native Space: Colonialism, Resistance and Reserves in British Columbia*, Syracuse University Press, Syracuse.

Howe, S. (2000), *Ireland and Empire. Colonial Legacies in Irish History and Culture*, Oxford University Press, Oxford.

Jones, A. and Jones, B. (2003), 'The Welsh World and the British Empire, c.1851-1939: An Exploration', *Journal of Imperial and Commonwealth History*, vol. 31, pp. 57-81.

Kennedy, D. (1996), 'Imperial History and Post-Colonial Theory', *Journal of Imperial and Commonwealth History*, vol. 24, pp. 345-63.

Lester, A. (2000), 'Historical Geographies of Imperialism', in B. Graham and C. Nash (eds), *Modern Historical Geographies*, Pearson, Harlow, pp. 100-120.

Lester, A. (2001), *Imperial Networks. Creating Identities in Nineteenth-Century South Africa and Britain*, Routledge, London.

Lester, A. (2002), 'Colonial Settlers and the Metropole: Racial Discourse in the Early 19th Century Cape Colony, Australia and New Zealand', *Landscape Research,* vol. 27, pp. 5-10.

Lloyd, D. (1999), *Ireland After History,* Cork University Press, Cork.

Loomba, A. (1998), *Colonialism/Postcolonialism,* Routledge, London.

Lowry, D. (2003), 'The Crown, Empire Loyalism and the Assimilation of Non-British White Subjects in the British World: An Argument Against 'Ethnic Determinism'', *Journal of Imperial and Commonwealth History,* vol. 31, pp. 96-120.

McLeod, J. (2000), *Beginning Postcolonialism,* Manchester University Press, Manchester.

Nash, C. (2002), 'Cultural Geography: Postcolonial Cultural Geographies', *Progress in Human Geography,* vol. 26, pp. 219-230.

O'Farrell, P. (1988), 'Landscapes of the Irish Immigrant Mind', in J. Hardy (ed.), *Stories of Australian Migration,* New South Wales University Press, Kensington, pp. 33-46.

Patterson, B. (ed.) (2002), *The Irish in New Zealand. Historical Contexts and Perspectives,* Stout Research Centre for New Zealand Studies, Wellington.

Philo, C. (2000), 'More Words, More Worlds. Reflections on the "Cultural Turn" and Human Geography', in I. Cook, D. Crouch, S. Naylor and J.R. Ryan (eds), *Cultural Turns/Geographical Turns: Perspectives on Cultural Geography,* Prentice Hall, Harlow, pp. 26-53.

Pierterse, J.N. and Parekh, B. (1995), 'Shifting Imaginaries: Decolonization, Internal Decolonization and Postcoloniality', in J.N. Pierterse and B. Parekh (eds), *The Decolonization of Imagination: Culture, Knowledge and Power,* Zed Books, London, pp. 1-19.

Pocock, J.G.A. (1975), British History: A Plea for a New Subject', *Journal of Modern History,* vol. 47, pp. 601-28.

Pocock, J.G.A. (1999), 'The New British History in Atlantic Perspective: An Antipodean Commentary', *American Historical Review,* vol. 104, pp. 490–500.

Prakash, G. (1996), 'Who's Afraid of Postcoloniality?', *Social Text,* vol. 49, pp. 187-203.

Prakash, G. (1999), *Another Reason: Science and the Imagination of Modern India,* Princeton University Press, Princeton.

Proudfoot, L. (2003), 'Landscape, Place and Memory: Towards a Geography of Irish Identities in Colonial Australia', in O. Walsh (ed.), *Ireland Abroad. Politics and Professions in the Nineteenth Century,* Four Courts Press, Dublin, pp. 172-85.

Ryan, J.R. (2004), 'Postcolonial Geographies', in J.S. Duncan, N.C. Johnson, and R.H. Schien (eds), *A Companion to Cultural Geography,* Blackwell, Oxford, pp. 469-84.

Said, E.W. (1978), *Orientalism,* Pantheon Books, New York.

Seddon, G. (1997), *Landprints: Reflections on Place and Landscape,* Cambridge University Press, Cambridge.

Sidaway, J.D. (2000), 'Postcolonial Geographies: An Exploratory Essay', *Progress in Human Geography,* vol. 24, pp. 591-612.

Smith, N. and Godlewska, A. (1994), 'Introduction: Critical Histories of Geography', in A. Godlewska and N. Smith (eds), *Geography and Empire,* Blackwell, Oxford, pp. 1-8.

Spivak, G.C. (1988), 'Can the Subaltern Speak?', in C. Nelson and L. Grossberg (eds), *Marxism and the Interpretation of Culture,* Macmillan, London, pp. 271-313.

Thomas, N. (1994), *Colonialism's Culture: Anthropology, Travel and Government,* Princeton University Press, Princeton.

Thompson, A.S. (2000), *Imperial Britain. The Empire in British Politics, c.1880-1932,* Longman, Harlow.

Trainor, L. (1994), *British Imperialism and Australian Nationalism,* Cambridge University Press, Cambridge.

Walia, S. (2001), *Edward Said and the Writing of History,* Icon Books, Cambridge.

Yeoh, B. (2000), 'Historical Geographies of the Colonised World', in B. Graham and C. Nash (eds), *Modern Historical Geographies*, Pearson, Harlow, pp. 146-66.

Young, R. (2001), *Postcolonialism: An Historical Introduction*, Blackwell, Oxford.

PART I
'DIS-LOCATIONS'

Chapter 2

Colonial Spaces and Sites of Resistance: Landed Estates in 19th Century Ireland

Paddy Duffy

Introduction

The regional expression of rural protest, agrarian outrage and rebellion in 19th century Ireland has been periodically examined by historians and historical geographers (Bric, 1985; Kiely and Nolan, 1992). One of the contexts within which such events may be re-visited is within a framework of local resistance to colonial domination. Post-colonial perspectives offer a critique of 19th century colonial discourse in which dominant power structures frequently served to 'Other' the colonized, through processes of negative stereotyping and myths of primitive backwardness. Colonial hegemonies generated both subservience and resistance in a variety of strategies by the colonized. Colonialism in its various manifestations throughout the British Empire provides some of the most clear-cut examples of a dominant elite subordinating a colonized 'inferior' native population.

Although the colonial nature of Ireland's relationships in the British Isles and British empire is sometimes ambiguous, they displayed most of the characteristics of colonial society in the 18th and 19th centuries, especially in the role played by the landed gentry. The Irish economy which was largely controlled in the 18th and 19th centuries by the English metropole, developed as an agricultural periphery to the British heartland. The Irish landed estate, which formed the lynch-pin of this economy, lay at the core of the colonial enterprise in Ireland from the 17th century. Estates became progressively contested spaces in the later 18th and 19th centuries, with increasingly unpopular attempts by the dominant elite to reform their properties through regulation. Ironically after 1801 Ireland was part of the United Kingdom and superficially less obviously a colony, though one commentator in 1834 noted the numbers of military garrisons in county Tipperary, as 'an array of bayonets that renders it difficult to believe that Ireland is other than a recently conquered territory, throughout which an enemy's army has just distributed its encampments' (McGrath, 1985). Much of the 19th century was taken up with agitation to remove the union between Great Britain and Ireland, generally in the teeth of opposition from the landed establishment which was predominantly unionist and imperialist in its outlook. The ambivalence and inherent contradictions in Ireland's political and colonial status are captured in the manner in which it was represented in the Great Exhibition in Dublin in 1853. The discursive logic of an exhibition that was designed to reconfigure Ireland symbolically as a modern progressive nation comfortably located in the United Kingdom and the Empire, in the end only served to emphasize its subordinate colonial status: the organizers of the Exhibition found themselves 'instructing people whom they consider as their own national lower orders in the behaviours appropriate to civilized life, by mobilizing colonial images of Ireland traditionally used to denigrate the island as a backward region of the United Kingdom' (Saris, 2000).

Many of the projects of settlement and survey which were undertaken in Ireland from the 17th century were essential components of colonial enterprise, measuring and mapping a conquered land for appropriation and domination. The construction of the landed estate system was largely a product of such a mapping enterprise. The Civil and Down Surveys in the 17th century, the Ordnance Survey and the General Valuation of the mid-19th century, for example, were all pioneering episodes which were later replicated in outreaches of the empire. Other elements of the 19th century Irish experience, such as the postal system, and the Irish Constabulary (Royal Irish Constabulary from 1867) which were crucial parts of the Irish administration, subsequently became models for colonial practice throughout the empire.

Much of the literature on postcolonialism, orientalism and the British Empire does not initially appear as overtly relevant to Ireland's case. Indeed Ireland itself, especially its landed elite, was involved in the consolidation of many imperial overseas projects. There are, however, parallels in British colonial experiments in 19th century empire and earlier colonial developments in 16th and 17th century Ireland, reflected for instance in plantations of settlers from Britain, transplantations and Cromwellian 'ethnic cleansing' of elements of the native population to the West Indies in the 1650s. Though the nature of colonial experience in Ireland was modified through the 18th century, it persisted in some of the attitudes to and treatment of the native Catholic population, with echoes of colonial domination throughout the 19th century. The Yahoos appear as an ironic satire on colonial perceptions of the native Irish in *Gulliver's Travels* (published in 1726), characteristic of many stereotyped 'others' in classic colonial discourse:

> ....the Yahoos appear to be the most unteachable of all Animals, their Capacities never reaching higher than to draw or carry Burthens ... For they are cunning, malicious, treacherous and revengeful. They are strong and hardy, but of a cowardly Spirit, and by Consequence insolent, abject, and cruel... the Red-haired of both Sexes are more libidinous and mischievous than the rest (Swift, 1953, p. 285).

Routledge talks of the 'place-specific' character of popular protest and struggles resulting from the manner in which society endows space and its associated resources with a variety of meanings (Routledge, 1997, 1997a). Land ownership and occupation as the ultimate expression of space relations is and has been the focus of tensions and the site of resistance between powerful elites and comparatively powerless landless people in many parts of the world. Ownership and control of land formed the fulcrum of colonial power in Ireland, with increasing proportions of immigrant (Protestant) landowners in the 17th and 18th centuries. The close interrelationship between the ascendancy/gentry and membership of the Anglican Church, British army garrison and Irish administration in Dublin demonstrates this. Lester has also underlined the 'critical spatial dimensions' of colonial discourse (Lester, 1998, p. 3). In Ireland estates can be seen as manifestations of such a discourse, particularly in the case of the more extensive properties of some thousands of acres. More than 95 per cent of the island's land resources were held by around 5000 landowning gentry in the 1770s, much of which incorporated comparatively large extents of territory over which one owner exercised considerable power and control. The owner of such an estate in early 19th century Ireland had 'infinitely more control over its inhabitants than the government ... having it in his power to render the little world of which he is the centre ... miserable or happy according to the principles of management pursued' (Thompson and Tierney, 1975, p. 83).

Such landed estates provide good examples of the operation and application of knowledge and power by dominant elites, reflected in networks of gentry intermarriage, visiting and correspondence within elite circuits, symbiotic linkages with the colonial administration in Dublin and London, estate agency theory and practice, all consolidated by close cultural affinities with England and its gentry elite. Lindsay Proudfoot has recently placed more emphasis on the hybrid nature of gentry identity within Ireland and less on its colonial status, suggesting that gentry/ascendancy elites were not unique to Ireland but were part of a wider European post-enlightenment age (Proudfoot, 2000; 2000a; 2001). It is impossible, however, to ignore the ultimately colonial nature of Ireland's experience within Europe, the reality of England's first colony on its western doorstep in which many of its more distant colonial endeavours were first tested, and its general perception by London as a troublesome colony throughout the 19th century. To Edmund Burke in the later 18th century, Ireland provided 'a metaphor for the world beyond Dover, affording points of comparison which helped to explain events in places as far-flung as India or the Americas' (Kiberd, 1995, p. 19). The ethnic/cultural constitution of Ireland's landed elite (who participated in the European enlightenment largely through the filter of a British gentry) adds a putative colonial dimension to its relations of power in Ireland and to its dominant-subordinate relationship with the mass of the tenant population. By the second half of the 19th century, it was becoming more and more a discredited, displaced and dispossessed elite.

Being very close to the heart of empire, there was an intensity and immediacy in impacts of colonial and imperial discourse on Ireland. The central role played by England in 19th century world capitalism meant that the social upheaval which all peasantries have undergone was experienced acutely by the Irish. This exacerbated the dislocations brought about by the modernization of the economy as Ireland was rapidly subsumed within the ambit of discourses of improvement and the new political economy pressuring traditional resources. Landed estates and their farms and townlands were the settings for what Scott has characterised as everyday local resistance, small expressions of dissent, disrespect and protest (Scott, 1990).

Estates as Colonial Space

In terms of origins, a great many of the landed estates were part of an overt colonial enterprize in the 16th and 17th centuries – involving confiscation, plantation and colonization of the land of Ireland by British settlers. Those whose lineages were not grounded in plantation policy (such as Anglo-Norman manors or Gaelic territories of the 16th and 17th centuries) were by the 18th century locked into a transparently colonial world to which they had largely conformed in cultural and religious terms. Whether the owners were English/British (like Devonshire, Lansdowne, Palmerston, Abercorn, Fitzwilliam, Bath or Shirley) or Irish/Anglo-Irish (such as Downshire, Leinster, Leslie, Charlemont, Fingall), they were overwhelmingly part of a class and system which reflected the consequences, and supported the project, of imperial hegemony in Ireland. Members of the ascendancy and gentry commonly sent their children to be educated in schools, universities and law schools in England or to the Anglican environment of Trinity College, Dublin (established by Elizabeth 1 in 1595), and served in the army (especially the Indian army), navy and colonial administrations overseas. 'Insecurity and the England-complex remained with them to the end' (Foster, 1988, p. 194); like many other British colonials, Anglo-Irish gentry like Elizabeth Bowen saw themselves as 'a hyphenated people, forever English in Ireland, forever Irish in England'... 'locked in misery between Holyhead and Kingstown [Dun Laoghaire],' (Kiberd, 1995, p. 367), a Kingstown whose placenames echoed with imperial ghosts – Wellington, Windsor, Carlyle.

Centres like Bath, Cowes and Cheltenham,[1] for example, comprised a familiar nexus of socialization for them. London and Dublin were foci of social, cultural and political life – though the Castle balls in Dublin by the 1850s were 'no great affair' in the lofty (English) opinion of Mrs Shirley.[2] Gentlemen's clubs, like the Carlton and Garrick in London or the Kildare Street Club in Dublin, were important components in Irish gentry networks. Many Irish landowners maintained a house in London in the 19th century: Sir Charles Powell Leslie's four daughters required 'a spring-board' in London in 1872 and Stratford House was duly leased (Dooley, 2001, p. 48). Evelyn Philip Shirley's daughter wrote to him in Monaghan from their London house in Belgrave Square in 1832 to inform him that the city was beginning 'to fill some coming to attend the Houses of Lords and Commons and others to be ready for the Levy and Drawing Room which take place this week, the Drawing Room on Friday to celebrate the Queen's Birthday'.[3] Like their British counterparts, Irish gentry went on Tours of the continent, Persia, Egypt, India, Africa, collecting arts and curios for their mansions. They married strictly within their class and creed, regularly seeking brides in England; London was 'the Mecca for matchmaking' (Leslie in Dooley, 2001, p. 67) and adhered generally to a collective view of the lower (Irish) classes, which generally represented them as 'other' to their morally and culturally superior world.

In much the same way as colonial authority elsewhere in empire was written into extravagant panoplies of landscape and architecture, the houses and landscapes of even the most modest gentry landowner in Ireland reflected a world of privileged extravagance, where the norms of 'civilization' and social order were inscribed in avenue and mansion, parkland, parterre and planted vistas – from the 'crenellated extravaganza' of Lord Gosford's castle in Armagh built during the 1820s to the Romantic landscape created by Frederick Trench at Heywood in Queen's county in the late 18th century (Thompson and Tierney, 1976, p. 8; Proudfoot, 2000a; Friel, 2000). There were also interior 'texts' in plasterwork, art collections and *de rigueur* trophies of the big game hunt: the Duke of Leinster collected speckled cows to ornament his demesne and shells from around the world for his Shell House; all aspired to fashionable Italian art collections. Here were statements in stone or plantation to reflect status and power, to impress neighbouring gentry and to instil deference and respect in the local tenantry: 'there is nothing will keep the Irish in their place like a well-appointed mansion' (Banville in Johnston, 1996, p. 556). Even small county towns managed to reflect a discourse of empire in their monuments and streetscapes, as stages for the politics of performance by landowning county society – marching troops, bands, and the flags and bunting of loyalty. The Dawson memorial in Monaghan commemorates Colonel Dawson's death in Inkerman in 1854. The town also contains a monument to Lord Rossmore and in the (Anglican) Church of Ireland there are memorials to the sons of gentry killed at Ferozeshah (India) in 1845 and Isandula (South Africa) in 1879 (Duffy, 2004).

These architectural and landscape statements were settings for a symbolism in day-to-day relations of power. Turlough House in Mayo for instance had the tenants step through great French doors into the Library to pay the rent, with a doffing of hats and appropriate gestures of subservience. Tenants lined the streets of Maynooth to bow to the Duke of Leinster *en route* to Sunday church. Gatelodgekeepers into the early 20th century bowed to the owners passing through (Somerville-Large, 1995). Evelyn John Shirley held annual tenant dinners in the Great Hall of his house at Carrickmacross at which lectures on frugality and industry were given to the tenants favoured with invitations. The moral and economic improvement of what was often seen as a sluggish and ill-disposed tenantry was the well-intentioned objective of the landed, reflected in one agent's remarks: 'when the beautiful variety of surface, which this country affords, is now observed bleak, dreary, and naked; and then look forward to it covered with well built cottages, well laid out farms, and thriving plantations, with contentment and its natural companions good order, peace, and prosperity reigning around, surely everyone ought to be tempted to put his hand to the work' (Blacker, 1837, p. 63, 66). Vaughan has suggested that there was more to improvements than agricultural innovation; 'They were the means by which landlords justified their existence, imposed their power on the countryside and enhanced their prestige' (Vaughan, 1994, p. 120). Foster's suggestion that they were attempting to legitimize their situation in Ireland more than a century after initial colonization has been comprehensively tested in a number of studies by Proudfoot (1993; 1997). A recent essay by Whelan emphasises the manner in which a colonial imagination is reflected in Anglo-Irish perceptions of ruins in Ireland as 'materialities of the colonized's defeat' and commemorations of the disorder of the pre-colonial world (Whelan, 2004).

Titles and patronage to accompany landed power and landscape display were eagerly sought out and highly regarded by gentry. In a gossipy letter to his son on the home estate in Warwickshire, Evelyn John Shirley reported on the attendance at a ball in his Carrickmacross mansion in 1848, paying special attention to the social status of the guests:

> Lord and Lady Farnham and two Miss Stapletons, Lord Worcester and Sir William Russell and Lt. Fraser of the 70th. ... Lord and Lady Fingall, and the Lady Plunkets and Lord Killeen, Lord and Lady Louth, Col Pratt and Mr and Mrs Chaloner and the Farrells, Ld Bellingham, Mrs Napier and two sons, Mr and Mrs Singleton came in a party of 21, and brought two beauties Miss Browns with their mother, and Mr and Mrs Coddington, two beautiful daughters etc, Major and Mrs McClintock, Mr and Mrs Olpherts, H Mitchell and daughter, Ruxtons, Longfields, Mrs Butler, Forsters (not Sir George or his daughter), Lucas's, Archdeacon Beresford and daughters, Lambarts, Winters, Smiths, G Filigate, C Fortescue, Sir P Leslie, Mr P Nicholson, Baronet Lestrange, Hawkshawe, Wooley, Proby, Folliott, Dillon, Lyle, Tipping, Wilman, Officers Capt Stewart, Lieuts Halfield, Coade, Hutchinson and Wassenbend of the 23rd Dragoons, etc etc.[4]

School, church, military and colonial service formed a network which cemented the solidarity of the landed elite. Towards the close of the 19th century, Castletownsend in west Cork was the setting for a close-knit community of a dozen Anglo-Irish families commemorated in many of the writings of Somerville and Ross: 'all Protestants, all suspicious of strangers and all completely sure of themselves ... Ireland continued to be divided neatly between US and THEM, and the ones who mattered in it, who made the place tick over at all, were undeniably US' (Fleming, in Scott, 2003, p. 1). Most of the influential personnel on larger estates, such as estate agents, agriculturists, stewards, clerks and other senior officials were usually recruited from the landed/colonial class, and commonly from Britain (Dooley, 2001). Grooms, gardeners, cooks and other 'loyal retainers' in Carton, in Kildare, for instance, were imported. The behaviour of Leslie's Irish footmen who accompanied the family to London for Queen Victoria's golden jubilee, got drunk and ran down Oxford Street shouting 'to hell with your bloody old Queen' probably confirmed the advisability of selecting employees who empathized with the world of Queen and empire! (Somerville-Large, 1995, p. 335). Cecil Frances Alexander, who married the Anglican bishop of Derry and Raphoe in 1850, was the author of the hymn 'All things bright and beautiful'. One of its verses might be seen as an important paean to imperial order, celebrated by the Established Church and its gentry adherents in the 19th century and sung throughout the empire:

> The rich man in his castle/The poor man at his gate
> God made them high and lowly/Each to his own estate.

Though there were tenurial constraints on land ownership in Ireland, especially in the 18th century, landed gentry were ultimately *the* centres of social, economic and political power and patronage from the 18th and into the 19th centuries. The more influential of them were frequently the objects of respectful dedication of work by authors, publishers, surveyors, architects and artists. Taylor and Skinner's map of County Louth, for instance, was presented 'with gratitude' to Jn Foster Knight of the shire of Louth. Newly published maps of Ireland, or its counties, in the late 18th century, pointedly represented the place of landlord and gentry elite in the developing Irish landscape, by marking the residences and mansion houses of gentry. In 18th and 19th century Ireland, as in England, the landed estate and its mansion was regularly used as a device in literary fiction to symbolise order and stability, especially a disorderly 'colonial' world like Ireland – from Edgeworth's *Castle Rackrent* at the end of the 18th century, to the works of Somerville and Ross in the late 19th century (Edgeworth, 1800; Somerville and Mount, 1920; Bowen, 1942). The society of the estate, privileged by religion and culture, continued to represent an exclusive colonial world-view in the 19th century which reluctantly gave way to the majority colonized/nationalist community in the dying years of that century. But into the early 20th century, the now dispossessed Irish landed gentry continued to hark back to an earlier world, before Elizabeth Bowen's 'golden close of the British 19th century'(Bowen in Scott, 2003, p. 26), with Lady Fingall writing that 'Irish landlords lived within their demesnes making a world of their own, with Ireland outside the gates' (Fingall in Somerville-Large, 1995, 355).

Irish landowner identities were complex and far from 'seamless' (Proudfoot, 1993). In addition, the personalities of individual landowners often emerge as a significant element in topographies of resistance at estate level – different owners having different management regimes, often dictated by different lifestyles and expenditures. Vaughan talks about the 'truculent meddling of Lord Leitrim', the 'paternal despotism' of Lord Fitzwilliam, 'hesitant fussiness' of the Gosford estate, the 'conscientious benevolence' of Hamilton, 'unbending integrity' of Mr Joly (Vaughan, 1994, p. 105). Moralistic paternalism, however, would characterize the generality of the landowners and associated gentry class (agents, clergy and military elites), with a growing penchant in 19th century for regulation and control. Regulation and stricter management of properties was more characteristic of landed estates in the post-war recessionary period from the 1820s, in sharp contrast to the more lackadaisical approach to estate management in the 18th century. William Steuart Trench, the land agent, writing for an English audience in the mid-19th century, noted that if the landlord 'ventures to interfere with old habits, old prejudices or old ways ... he must be prepared to contend with difficulties which none but those who have experienced them could have imagined' (Trench, 1868, p. vii). On the other hand, absentee landlords, like Sir William Palmer with lands in Mayo, who took little interest in their estates and the tenants, left all to agents whose main focus was to get in the rent at all costs: the result was mistrust and antipathy between owners and occupiers.

Throughout the country there were innumerable instances of the estate as the arbiter of power and privilege, and a demonstration of 'civilization' and 'superior moral order' in operation. The law of the land was mediated through the estate's representatives: resident magistrates and justices of the peace were usually landowners or their agents and most landowners had ready access to and contacts in the centre of Irish administration in Dublin Castle – like Sir William Palmer who wrote (from Wales) to the Castle in 1847 seeking the erection of a police barracks on part of his Mayo estates to assist his agent in the collection of his rent (Byrne, 1996). County Grand Juries were composed of the propertied class who administered justice and local government. Throughout the 18th and 19th centuries, political representation (either in Dublin or Westminster) was seen as the automatic entitlement of the landowning classes. Networks of patronage emanating from Dublin Castle lent authority and influence to them, as well as ensuring their support: a typical government memorandum book of 1818 recorded details of patronage in Westmeath – 'Lieut. Colonel the Hon. H. Pakenham Lord Longford is a Representative Peer; is Custos Rotolorum of the County; is a Trustee of the Linen Board. He is brother-in-law of the Duke of Wellington. His brother, Admiral Thomas Pakenham has £1200 compensation as Master General of the Ordnance' (Jupp, 1973, p. 166). Charles Powell Leslie (also related to the Duke of Wellington), on losing the 1826 election in Monaghan to the pro-Catholic Henry Westenra, thanked his supporters and hoped that they would 'uphold what the state of your country, as well as that of the empire demands, the Protestant ascendancy in church and state'.[5]

Networks of patronage were also replicated at local level: William George Smith, clerk in the Shirley estate office in Carrickmacross sought a 61-year lease in 1839 from the agent for some property on the estate pointing to the benefits of favouring him:

> by making this purchase, I have…secured to myself and representatives after me the means of furthering in every possible way, the interest of Mr Shirley and his political friends, … Should Mr Shirley think it right to consider this favour to me, … Protestants having capital may by this example be encouraged to bring it to bear on the removal of his opponents and a class of persons might thus be induced to settle on his property who … might hold in check, not only his enemies but those of the established institutions of the country.[6]

We can talk, therefore, of an estate *system* that was implicated in many of the predilections of colonialism – 'order', 'improvement', 'civilization', integrity, morality, industry, loyalty, subservience. Shane Leslie, author and landlord, in his novel *Doomsland*, likened the running of the estate to an old watermill: its business proceeded routinely from season to season: 'a hundred cogs moved and clicked in their place... Agents, clerks, land-stewards, bog-bailiffs, gardeners, gamekeepers carried out dilatory functions. Wages, jointures, salaries, tithes, taxes, pensions, mortgages were paid' (Leslie, 1923, p. 25). Brenda Yeoh (2000) has documented the manner in which colonial governance in Singapore regulated and regimented society and space to produce a western British expression of municipal order. Landownership in Ireland saw its role in much the same way as the colonial administration in India for instance, whose duty was 'to impose linearity and order on an ungovernable society' (Chatterjee, 2000, p. 20). It was supported by an increasingly bureaucratic State system which, up until the Great Famine in the late 1840s, largely unquestioningly supported the landowning elite in Ireland, manifested through networks of patronage and political preferment.

As disturbances and resistance increased in a variety of regions in the 19th century, the police and military forces of the state were expected to support the landed establishment. This support also traditionally extended to the Anglican (and what might be characterized as the colonial) church in its collection of tithes: proctors and process servers were backed up by parties of police and militia during the 'tithe war' in 1830s. Fourteen people were killed in Bunclody in County Wexford during tithe agitation in 1831, and 11 policemen were killed in Kilkenny disturbances in 1832 (Kieley and Nolan, 1992). Evictions near Woodford in Galway in 1843 were assisted by a force of 300 of the 5th Fusiliers, a troop of 4th Royal Irish Dragoons, a troop of 10th Hussars, and 200 policemen (Clark, 1979, p. 69). Thomas Drummond, who was appointed Under-Secretary for Ireland in the late 1830s, was a reformer who began the questioning of this alliance of landed elite and state with his observation on property having duties as well as rights. Only gradually, as the 'land war' intensified after the Famine did the State contemplate withdrawing such unconditional support for Irish gentry.

Perspectives on the Colonized

Although like the landowning elite, the largely Catholic farming class was also far from being a seamless community, in general, gentry perspectives tended to simplify and collapse all into a collective colonially-inferior status, usually tinted by anti Catholic prejudice. Into the late 19th century, the gentry had a real horror of intermarriage with Catholics of any hue, reflected in the novels of Somerville and Ross. Indeed the marriage of Shirley's younger son to a Dublin Catholic in the 1850s precipitated a family crisis. Looks, accent, dress and general deportment were seen as important markers feeding into a universal representation of peasantry as other, characterized in recurring terminologies of 'ill-disposed', 'slothful', 'wild', 'wily', 'cunning', feckless etc. Trench's favourite metaphor for managing tenantry was 'harness,' to restrain a population, which he characterized as 'docile and easily led', and generally obedient to their superiors, yet 'when once assembled in masses .... they become capable of the wildest and most frenzied excitement' (Trench, 1869, 70).

Even in the late-19th century, some of the writings of Somerville and Ross depicted networks of gentry houses sprinkled through landscapes which were the settings for hunts and elite outdoor pursuits and were peopled by an array of tenantry, distinguished by subservience, humour, slyness, 'blarney' – idiosyncratic 'difference' looked down on from the saddles of the Galway Blazers or the Rosscarbery Hunt (e.g. Somerville and Ross, 1901). Social distanciation helped to reduce individuals to a uniform mass, invariably seen through a coach window, from horseback, or through the colonial lens of the press. Thomas Carlyle was confirmed in all his imperial prejudices about the Irish during his Famine visit. From the upper platform of the mail boat to Ireland, he observed five or six typical degenerated 'physiognomies'– one 'a lean withered slave of a creature with hairy brows, droop nose, mouth corners drooping, chin narrow, eyes full of sorrow and rage', all 'with the air of faculty misbred and gone to waste' (Carlyle from 1882 in Crowley, 2003, p. 163).[7]

Trench while agent on the Lansdowne estate was captivated by a peasant girl in terms which have many of the characterizations of 'Othering', justifying her attraction to him by separating her from his class view of the local peasantry – putting her in an altogether more acceptable category:

> She had but little of the original Celt in her features. Her beauty was purely Spanish, of which I have seen many perfect specimens in Tuosist and around Kenmare: large soft eyes, with beautiful dark downy eyelashes, the mouth well formed, and cheek of classic mould ... The form which now stood before me was a beautiful specimen of this perfect Spanish type ... her hands were clasped in an attitude of wild supplication ... she was perfectly natural and simple, and ... so intelligent a girl as she was could not possibly look at her reflection in one of her own dark mountain lakes, and not see that she was different from her neighbours ... She had watched my countenance with the quickness of an Irish peasant during the whole time she was speaking (Trench, 1868, pp. 76-77).

E.J. Shirley in the mid 19th century held many of these opinions of his Irish tenants. Like many others in Ireland, he was an English landowner with an estate in Warwickshire; his overcrowded and sprawling Irish property with its 20,000 population must have appeared to him as a colonial outpost, many of whose inhabitants he sometimes considered a wild and fickle rabble. On the other hand, it also had an exotic attraction which drew him every summer from the late 1820s, and he commenced to build an extravagant mock Tudor mansion in the 1830s. Like Lord Farnham of Cavan, Lord Roden in Down and many others, Shirley saw himself as superior guardian of his tenantry and the moralistic paternalism of his many utterings were regularly published on handbills and addresses for dispersal throughout his estate, reflecting an outlook expressed by Lord Lansdowne in 1870: 'the longer I live the more firmly do I believe in blood and breeding' (Lyne, 2001, p. xliii). One of the law officers on the Shirley estate probably fairly represented the gentry perception of the mass of the tenantry at the height of a rent strike in 1843:

> people are still hanging back and *skulking* behind the *pretension* of danger to themselves or property if they pay their rent; ... but this feeling has its origins in the *baseness of character* so very prevalent in the *absence of right moral principle* ...' [emphasis added][8]

Ignorance, sloth and cunning were common traits attributed to Irish tenants by their superiors in the 19th century and may be illustrated by excerpts from private correspondence between members of the landowning gentry, reflecting an essentially internal discourse not intended for public consumption. Writing to Lord Wilburton in 1850, in relation to the management of the neighbouring Bath estate, Shirley suggested that '*vigorous measures* to obtain rent .... are *absolutely necessary* in dealing with the "Celts of Farney", most of whom only pay by compulsion ... The people are very quick and cunning, ... many of the tenants are idle and reckless and in some instances neglect all improvement.'[9] Trench in considering the offer of the Bath agency wrote about the principles which he adhered to in running estates: that he would 'place these wild and uncivilized people sufficiently under his command that he can force them by a judicious mixture of firmness and kindness with the duties and responsibilities of some approach to civilized life.'[10] He further emphasized this approach a month later, significantly manifesting some fundamentals of colonial discourse: 'nothing impresses a wild though intelligent and cunning people with more confidence in a superior, than his at once showing them that he is accurately acquainted with the whole subject in hand and nothing tends to anarchy and confusion as even the *appearance* of indecision or ignorance'.[11] Thomas Derby an English land agent in Tipperary wrote to Trench about the necessity of assisting emigration off landed estates: 'so as to strip off the *Rubbish* (may God forgive the word, but I speak only in the way of utter want of intelligence, industry and hopelessness of improvement) and preserve a sufficiency of stock of an improvable kind.'[12]

The Ordnance Survey Memoirs, written in most cases by English military officers, contain comprehensive and valuable accounts of rural economy and society in the north of Ireland in the 1830s. They often betray the external and elite background of the authors in their frequent representations of local communities as primitive and indolent, indicting them for their tendencies to abandon the harvest for a wake or a wedding or fair. 'Would that the habits of industry and the desire of accumulation, which so generally prevails in the sister kingdoms, could be imparted effectually into the minds and disposition of this wretched and deplorable peasantry'......'A glance at the wretched hovels, scantily covered with straw, surrounded and almost entombed in mire, which everywhere present themselves ... sufficiently testify that the total absence of all activity in industry is one source of the wretchedness and misery which almost overwhelms the land' (Day and McWilliams, 1998, Parish of Laragh 44; Parish of Currin 93). Carlyle searched (Europe's) imperial outposts for landscapes analogous to what he saw in the west of Ireland: 'Claddagh as like Madagascar as England. A kind of charm in that poor savage freedom' (Crowley, 2003, p. 165). Outside observers were universally impressed with 'the laziness and idleness' of estate tenants, the 'vivacity of the Celt and a portion of ... sloth and cunning'.[13]

Shirley's annual addresses to his tenants prior to his return to his Warwickshire estate were imbued with well-meaning paternalism which advocated, for example, 'an encreased degree of *improvement, comfort and respectability*. To promote this and to encourage a love of *order, tidiness and cleanliness* is the anxious desire of my heart' and deprecated 'the scenes of drunkenness too often exhibited in market and fair days in Carrickmacross'.[14]

He was concerned with the general indolence of his tenants:

Now it is my duty to tell you that you do not value your time...I saw numbers of the tenantry lounging and idling about, their turf still in the Bog, their Hay still uncut, and the weeds growing plentifully... Pro. ch XV, verse 19, 'The way of the slothful man is as a hedge of thorns'[15]

Like many of his peers, Shirley planned to appoint a Moral Agent (who was an English army officer) for the estate in 1839. Lord Farnham in Cavan had his estate divided into districts in 1830 to facilitate its 'moral management'. William Krause, born in the West Indies but living in England, was his Moral Agent from 1826 and 1838, whose duty it was to 'free Roman Catholics from bondage'. The lives of the tenants were closely monitored by him and illicit distilling of poitín was prohibited on pain of eviction, while bawdy ballads and vices such as swearing, gambling and dancing were forbidden. Evangelical Protestantism was instinctively linked with social and economic progress, in the same way that Irish Catholic culture was associated with backwardness and inefficiency (Hill, 2002, pp. 78-79). A similar concern with moral welfare drove Trench, while agent on the Lansdowne estate in Kerry, to eject a tenant who had moved in with the daughter of a neighbour and married her a year later: as well as contravening the marriage law of estate, they were condemned because they had 'committed fornication together'.[16]

The Marquis of Downshire kept himself well informed on his tenants' behaviour, frequently engaging in tours of inspection: 'Wilson the Master of B.macbrennan School does not do enough for his Wages....he seems sleepy and has few Scholars, Schoolroom dirty and full of dust and very untidy.' 'Two persons named Orr live in a wretched hole with four acres .... The man came out to me without his breeches. They should not hold land' (Maguire, 1974, p. 141). In the age of Malthus and Darwin, natural science provided Downshire and others with ready metaphors for lessons on moral and economic improvement, with references to weeds and corn being suitably biblical in tone: 'Luxuriant as is the Growth of Good Crops, as well as noxious Weeds, the plant Truth is of dubious growth & its offspring, plain dealing and openness are often stifled in the Birth' (Maguire, 1974, pp. 153-54). Trench and others frequently reached for animal analogies to depict the state and character of many of their tenants – 'superabundant rabbits' .... 'locusts,' whose campaign of protest against the Lansdowne estate was dampened by cold and wet weather by which they were 'daily driven into their burrows' (Lyne, 2001, p. 301).

Regulated Space

Power and authority produces and regulates space and place and its occupants, surveying and territorializing it, and imposing discipline through, for instance, controlling movement within it. Space thus becomes the setting and catalyst within which authority may be challenged or resisted. One of the great preoccupations of estate management in the 19th century, especially in the years after the end of the European war, arose from pressure on land and questions of population control, management and regulation of access to land. This formed a universal context for estates as contested landscapes in 19th century Ireland. Although for a great proportion of the peasantry and small tenants, it was the larger (middleman) farmer who sublet land to them, the 19th century saw the gentry owners of the land intervening to reorganize and re-establish order in landholding relations.

From the perspective of landowners and land agents, regulations were aimed at rent control and payment, shortening leases, subdivision and subletting of farms, restrictions or control of access to other local environmental resources (such as woodland, bogland, rivers, lakes, mountain), sale and transfer of farms, improvements in farms such as enclosures and squaring of fields, improvement in houses, as well in some cases as the education and general moral behaviour of the tenantry. At local estate level, manor courts frequently operated to implement estate rules and regulations, and fines and punishments designed to order society and landscape within its boundaries.

Most of the proactive intervention took place as early as the mid-18th century on the best land, with some belated interest in the marginal western lands in post-famine years. Estates with the most active intervention in lives and landscapes of their tenants reaped a legacy of hostility – especially in post 1820s and were subsequently execrated in folk memory. Agents like Mitchell, Trench and Morant on the Shirley estate, or Lord Palmer or Lord Lucan in Mayo enforced estate rules and earned a reputation as exterminating landlords. It was Trench's somewhat exaggerated belief that the 'careless, spendthrift, good-for-nothing landlord, who hunts and shoots, and drinks and runs into debt, who ever exacts the most exorbitant rents from his tenants, provided only he does not interfere with their time-honoured customs of subdividing, squatting ... and reckless marriages, may live in peace ... in high favour with the surrounding peasantry' (Trench, 1868, p. 47).

Rural population growth expanded rapidly in the poorer regions where there had been little 'colonial' interest or commercial incentive to manage properly and on less supervised spaces such as back lanes, bog edges or roadsides where squatting prevailed: on the Midleton estate in Cork in the 1840s, for example, a subtenant of 37 acres had allowed 45 cabins on the side of the road; on another 60 acre farm, the tenant had allowed 98 cabins to be 'thrown up' on the sides of the road (Donnelly, 1975, p. 13). Shirley's 26,000 acre estate in Monaghan had a population of some 20,000 in 1841 – approximately 3000 tenants and 600 cottiers. Palmer's 80,000 acres in Mayo supported 23,000 in 1841. Lord Lansdowne's 96,000 acres in Kerry had a population of approximately 17,000. In these regions and places, landowners who were by inclination and politics doctrinaire Malthusians by the 1820s, were keen on relieving their properties of the burdens of excessive numbers of people.

'The tenants big and little are too much in the habit of subdividing their holdings, selling and setting, chopping and changing as their will leads them, defying me and my regulations' (Maguire, 1974, p. 142). Subdivision of farms among family members and subletting portions of farms to cottiers were prohibited outright on most landed estates. The Devon Commission of enquiry into agrarian conditions in Ireland (1843-45) was heavily preoccupied with the enforcement of rules about subdivision, which even on the most well managed estates were more honoured in the breach than the observance. Most leases had clauses prohibiting these practices on pain of eviction but until the 1820s and 1830s strict enforcement was rare: references to 'nests' and 'swarms' of cottiers are regular occurrences in contemporary estate correspondence. But by the mid-19th century a great many landowners were unable or unwilling to provoke the hostility of the local population by intervening in cases of subdivision. The 'cold fear of provoking agrarian outrage' restrained many landlords in their regulation of tenantry (Donnelly, 1975, p. 54). In many cases it was difficult to police the enormous populations on the larger estates. Advisors like Blacker and Trench suggested that estate managers undertake thorough surveys of their properties including details on the circumstances of each tenant and the condition of their farms, houses, offices and stock, as well as the numbers and ages of their children. Regular reviews would help to 'distinguish and encourage the deserving' (Blacker, 1837, p. 4).

The links between tenant marriages, demographic growth and farm structures were well known to have repercussions on the future well-being of estates. Subdivision of farms, for instance, inevitably followed marriages of family members and many proprietors assumed rights to oversee the marriages of their tenants. Elizabeth Smith, lately arrived from Scotland, was aghast in 1839 at the careless manner in which tenants in Ireland married without reference to their landlord, frequently setting up home in outbuildings (Trant, 1997, p. 49). Shirley in the 1830s ineffectually admonished his tenants for subdividing ... 'abstain from leaving in your wills what is not yours to leave', and on both the Shirley and Bath estates marriage regulations, by which licences had to be obtained from the agent before going to the clergyman, were in place up to the 1860s. In 1842 Shirley had a handbill posted throughout his estate addressing the marriage issue in strident Malthusian terms:

> The necessity of consideration before engaging in marriage is self-evident.... remember that you injure your neighbours by throwing upon them the burthen of supporting those whom you ought yourself to support.....taking employment and food from those who already have not enough...Keep animal impulses under the control of reason.[17]

Lord Palmer also enforced rules regarding marriage and families sharing houses in the post Famine period, in one case demolishing a house and confiscating crops in 1864 for contravention of the rules (Byrne, 1996, p. 61). From the Famine onwards there was growing local resentment and resistance to these interventionist policies.

The organization of surveillance and information networks on estates became increasingly repressive in the 19th century. Local informants such as 'keepers' and 'watchers' were responsible for informing the estate on the progress of tenants' crops and other activities, preventing some from selling off at harvest time and absconding without paying rent, as well as searching the premises of defaulting tenants. Tenants were alert to the presence among them (in chapel congregations, for instance) of landlord-favoured tenants, employees or others such as gamekeepers and Royal Irish Constabulary men who acted as the 'eyes and ears' of the estate. The Bath tenants petitioned the agent in 1849 for the abolition of the 'odious system of placing keepers on their property'.[18] Many estates with absentee owners, however, such as Crown estates, did not have as effective a knowledge of their tenants' affairs (Scally, 1995).

Bailiffs, grippers and process servers applied the rules of the estate, driving off the livestock of defaulting tenants or making arrests. Trench, land agent at different times on the Shirley, Lansdowne and Bath estates, made it his business to become acquainted with every aspect of the estate he was on and regularly boasted in the 1850s that: 'a mouse can hardly move [or a dog cannot bark] on the estate without my knowledge.' Indeed Trench himself had earlier considered the application of some of the regulations in 1843 to be rather too rigorous on the Shirley estate, provoking a rent strike and other disturbances. He highlighted a range of penalties all aimed to keep the tenants, as he put it, 'tightly to their traces'. Decrees were taken out at the quarter sessions against defaulting tenants, and if the rent was not paid promptly, arrest and imprisonment followed, with 'ruinous expenses and loss', ensuing for the tenant.[19] This period of increasing imposition of regulations and reform on estates coincided with post-war depression, rising arrears, falling rents and Poor Law taxation increasing pressure on tenants. And it resulted in rising opposition and resistance, with sporadic and largely uncoordinated outbreaks of violence, burnings and maiming of cattle being extensive in the 1840s. Regular 'outrage reports' were made to the government by the constabulary from the counties concerned.

With the intensification of modernization processes, the 19th century thus witnessed a collision of the small tenant farmers' 'moral economy' and its traditional and customary relations with the land, with the expanding market economy of Britain into which the landed estates were locked. Restrictions on such ancillary resources as turbary or woodland, with fines for what Trench referred to as 'misconduct and disobedience' (see also Lyne, 2001, p. 262) became common, as estates determined to extract maximum value for all aspects of the property. The attitude of management to goats, for example, regarded as an animal of the poor and marginal, sums up the clash between poverty and improvement. The capacity of goats to eat newly-planted hedges, usually given gratis by interested landowners as an encouragement to tenants to enclose newly-squared fields, led to regulations to restrict them. Lord Leitrim, Shirley and others, prohibited many of their tenants from keeping goats. These restrictions of marginal economic activities of tenants were regarded with hostility by tenants and Shirley was still remembered as a 'bad landlord' by schoolchildren in 1938 for this reason.[20]

Economic change which followed colonialism in southeast Asia in the early 20th century had precedents in 18th- and 19th-centuries Ireland, where all the resources of the land which had traditionally been part of communal ownership – commonages, wastelands, woodlands, fisheries, and in Ireland's case turbary and rundale farming in the western regions – were appropriated by landlord estates, whose property rights were enforced by the colonial state's militia and courts (Scott, 1976; Kiely and Nolan, 1992, pp. 472-474). In South East Asia, as in Ireland a century earlier, this inequality in access to land resources was exacerbated by rural population growth leading to everyday forms of resistance organised on a kinship and territorial basis – boycotting, strikes, assaults, rioting, destruction of property (Brass, 2000, p. 133; Clark and Donnelly, 1983, p. 7). Modernization of agriculture on Irish estates in the 19th century involved improvements designed to eliminate joint tenancies, rundale settlements, consolidation of tenant holdings, prevention of subletting, restriction of access to turbary, woodlands, game, generating resistance by tenants which echoed agitation and repeated the practices of earlier generations of rural protesters: 'we, levellers and avengers for the wrongs done to the poor, have unanimously assembled to raze walls and ditches that have been made to enclose commons' (Bric, 1985, p. 153).

Estate management regulations were devised to compel acquiescence by a sometimes recalcitrant tenantry and formed the context for resistance on more and more estates during the mid-19th century. Such growing resistance meant that estates had to be judicious in dealings with tenants, lest as Trench expressed it in terms of traces and harness, 'that they must needs go steadily forward, or else by some violent plunge break through all restraint' (Duffy, 1997, p. 116). By the 1870s, tenant resistance in much of the country disabled many management initiatives: on one estate in Limerick the bailiff refused to serve notices to quit in 1878 because he was afraid that the tenants 'would do away with him' (Donnelly, 1975, p. 196).

Strategies of Resistance

Most features of agitation and protest evident on Irish landed estates reflect closely the various stratagems of subordinate resistance examined by Scott (1990). The public performance of subservience, putting on a show of humility, masking true feelings or flattering to deceive as rituals of subordination are characteristic responses by the comparatively powerless, 'the colonized native who understands that because his security depends upon compliance with the system he needs to display total loyalty' (McLoughlin, 1999, p. 202). And this confirmed elite perceptions of the subordinate other/natives as innately deceitful, evasive, fickle, and cunning. In this way what was characterized as 'oriental inscrutability' made the real Burman discourse inaccessible to the British (Scott, 1990, p. 35). (Similarly the intentions and attitudes of the colonial world, as exposed in the private correspondence of the Irish landowning elite, for instance, were equally unknown to the local community.) Disguise and anonymity, therefore, are the hallmarks of subordinate protest: use of anonymous threats and intimidation in circumstances 'where any open, identified resistance to the ruling power may result in instant retaliation', means that the regulations of the powerful can most effectively be countered by the anonymous threat of violence intended, for example, to 'chill the spine of gentry, magistrates' (Thompson in Scott, 1990, pp. 148-149). Coded signals of resistance, especially in songs, ballads and folktales, where the real object of protest is mocked with irony or satire, represent another form of hidden resistance which has been used in Ireland. Universally, pressure by dominant classes on traditional rights with restrictions on local access to resources, are countered in a range of 'down-to-earth, low-profile stratagems designed to minimize appropriation' (Thompson in Scott, 1990, p. 188; see also Brass, 2000, p. 130): theft, pilfering, shirking, evasion, foot dragging, sabotage of crops and livestock, arson, flight, poaching, squatting, beating gamekeepers and other representatives of authority, for example. In the eyes of the dominant/colonial elite, many of these stratagems of resistance are not the consequences of application of arbitrary power but 'of the inborn characteristics of the subordinate group itself' – who are by nature lazy, lying and unreliable (Scott, 1990, p. 37).

Passive resistance in 19th century Ireland was most classically exemplified in the 'boycott', which characterised the climax of the breakdown in gentry-tenant relations during the Land League disturbances in 1880. Captain Boycott was the land agent on Lord Erne's estate in county Mayo who was ostracized by the local tenantry ('boycotted') during a rent dispute. He imported some hundreds of Orangemen from south Ulster to assist with his harvest, guarded by some 1000 troops in an ultimately futile demonstration of landlord power. 'Combinations' such as this were a flagrant challenge to the authority of the landowning establishment: 'it is always the spectre of an open rebellion by the peasantry which haunts the conscience of the dominant classes in agrarian societies and shapes their exercise of domination' (Chatterjee, 2000, p. 22).

Between 1800 and the Great Famine, the government passed 35 Coercion Acts to control lawlessness in Ireland, mostly collective violence in rural areas (Clark, 1979, pp. 66-67). In 1827 for example, magistrates in Tipperary petitioned the government on two occasions to implement the Insurrection Act which imposed curfew from sunset to dawn (McGrath, 1985, p. 275). The estate which increasingly controlled and restricted access to land was at the coalface of these local resistance strategies. They ranged from passive non-cooperation, through anonymous night-time protests, rent strikes, intimidation of estate employees, to the murder of landlords and agents. Earlier practices of protest were resurrected – cropping of horses and maiming of cattle, destroying crops, firing shots, attacking and burning houses, levelling fences, burning turf stacks, administering oaths to secret societies, as well as erecting gallows, digging open graves, and despatching threatening letters and placards.

Protests on the Shirley and Bath estates in the middle decades of the 19th century involved many of these kinds of tenant resistance to the management regime on these two extensive properties. The agents on both estates were long aware of the potential for tenant protest, as far back as 1795 warning that to guard against a 'combining disposition in the tenantry', leases should not be allowed to expire at the same time over the two estates.[21] There was a constant awareness on the part of the estate administration in the mid-19th century of the presence of a stratum of insubordinate tenantry: in 1850 Shirley's visit to the agricultural show was a measure of his (un)popularity – 'some tried to give me a *cheer* on going into the show ground which was so poor an attempt I may say it failed. There were not enough Tenants and the *rabble* of course did not care about joining'.[22] Trench reported to Lord Bath in 1853 on the 'mob of reckless paupers' on the estate.[23]

In the 1830s, Shirley's commitment to improving the general education of the tenantry was evident in his public exhortations to improve their moral and social behaviour. With the help of the Kildare Street Society, which was actively committed to proselytism, he established a number of schools, which emphasized bible reading, on his estate:

the word of God is important. For this end I have provided for the Protestant the authorised and for the Roman Catholic youth, the Rhemish version ..... I hope the time is not distant when the native Irishman will be brought to obey the divine command to 'search the scriptures' and to consider it his right as a human being, his duty as a Christian and his *privilege* as a *British subject*.[24]

The reaction by the tenants demonstrated the effectiveness of 'combination' in protest, as well as the role of the priest as leader. The Roman Catholic clergy were opposed to what was perceived as proselytism by the estate. One of the schools, held in a Catholic sacristy by the clerk of the chapel who was paid a salary by the estate agent, was abandoned by most tenants. Schoolmaster and priest brought their case to the petty sessions in Carrickmacross where both magistrates, agents of the Shirley and Bath estates, supported the continuance of the school. In consequence, the local community, in the words of the priest, 'took the law into their own hands, came at night in a body and levelled the sacristy to its foundations...scattering to the winds all the bible and proseletyzing tracts' (O'Mearáin, 1981, p. 409). Several other estate schools were attacked in similar fashion and their (predominantly Catholic) teachers beaten. Such demolitions were not uncommon in other parts of the country in the 19th century. In 1839 when the parish obtained a grant to establish a national school in the chapel yard, the land agent initially prohibited the supply of building materials from any part of the Shirley estate. Neighbouring parishes, however, provided assistance with a convoy of carts during the night and, in the words of the parish priest, 'on the following day as much stones, sand and lime were left on the ground as built the schoolhouse' (O'Mearáin, 1981, p. 409). Soon after, Shirley cooperated in the granting of sites for a number of national schools on the estate.

Alexander Mitchell's agency of the Shirley estate from 1830 was accompanied by the gradual tightening up of management and control of the property. Rents were enforced by impounding the livestock of defaulters, or by putting them in gaol. Fuel rights from turf bogs were an increasing concern to Irish landowners especially where cottier populations had escalated. Bog rents were imposed on the Shirley estate in the early thirties and bog tickets had to be purchased in the estate office. Furthermore, lime rents for limestone burnt in kilns on the estate were enforced. 'Raising the coppers' was a book-keeping practice adopted by the office in which rents and other charges to the tenants were rounded up to the shilling. Both practices Trench later agreed, represented 'close shaving' by the estate to boost income, which only served to aggravate the tenants (Duffy, 1997, p. 114).

These impositions were the cause of an outbreak of protests in 1843 following Mitchell's sudden death. His death was signalled by a rash of celebratory bonfires on the hills of the estate. Bonfires were universal signals of dissent and protest in many parts of rural Ireland. A large public demonstration was arranged to meet the new agent William Steuart Trench, demanding a reduction in rent: 'down with the coppers', 'we'll stand the grippers no longer', 'we'll hang the keepers' (Trench, 1869, p. 79). In April and May of 1843 the disturbances continued on the estate with a rent strike and bog protest by the tenants. Anonymous placarding took place throughout the estate, the police informing the office that one had been posted on the chapels: 'On Tuesday the ninth of May let each person go to his Bog to cut his Turf, it is the advice of John Lattitat (sic) – let there be a water pool ready for the bog trotter'.[25] Latitat was a legal device, bog trotter was the bog ranger. Anonymous placarding was a popular medium of protest on the Lansdowne estate in the 1850s as well as in many other estates (Lyne, 2001, pp. 294-296; Kiely and Nolan, 1992, p. 468).

The estate retaliated vigorously with notices to quit and impoundment of the cattle of defaulters. Shirley found himself the centre of national attention in defending landlord rights. On application to Dublin Castle, a troop of horse and company of infantry were despatched to Carrickmacross to support the estate. Attempts to post ejectment notices on a chapel in the estate by estate officials, accompanied by police and military, were signalled by bonfires and accompanied by jeering crowds. More troops were sent from Carrickmacross, the Riot Act was read, and soldiers opened fire killing a tenant.[26] Driving in the cattle of rent defaulters, accompanied by police, bailiffs, the agent and three or four magistrates turned into a farce, as Trench described it in his memoirs: 'Not a hoof nor a horn was left in the countryside', as all the cattle had been spirited away by the tenants, and a forlorn little heifer was all that they succeeded in driving to the pound to the 'jeers and laughter of the populace' (Trench, 1869, p. 85).

In the following months, the agitation escalated with violent attacks on bailiffs and other officials connected with the estate. Drivers were threatened in anonymous posters: 'We will dissect you alive – life is sweet'.[27] At night the Molly Maguires took to the roads compelling the support of the tenantry and intimidating bailiffs and drivers. In Trench's words, the Molly Maguires were:

> stout active young men, dressed up in women's clothing, with faces blackened; or otherwise disguised; sometimes they wore crape over their countenances, sometimes they smeared themselves in the most fantastic manner with burnt cork ... to suddenly surprise the unfortunate grippers, keepers or process-servers, and either duck them in bog-holes, or beat them in the most unmerciful manner (Trench, 1869, p. 85).

The terror imposed by such groups, often recruited from neighbouring estates, is well reflected in a plaintive letter to the estate from a tenant seeking arms for his defence against what he called the Bundoran Girls:

> I take the liberty to inform you of the dangerous state I am in. On Monday I went down to Coraghy to see Mr Shirley's house, as I was proceeding home I espied ... some women looking earnestly at me one of them started up the road before me ... When she came some distance, I saw a number of women standing along with her. But they were men in women's clothes. Were it not how I proceeded in haste through the country I really believe I would have been murdered for the Bundoran girls were marching through Cornenty the same day. The reason I would be beat is it is reported that I am one of your Honor's bog bailiffs.[28]

In the end, on Trench's advice, Shirley conceded most of the demands of the tenants – many of the impositions and charges were removed or reduced.

In 1849 Shirley embarked on a series of evictions which attracted the attention of the *Nation* newspaper and became a *cause célèbre* in Britain and Ireland. Offers of passage to America were made to some of these tenants, but the evictions were frequently resisted by barricades and pailfuls of boiling water thrown at bailiffs from inside the houses.[29] In these post-famine years, as pressure was exerted on the tenantry by the landowners, agitation and intimidation spread throughout the region with the result that the county was proclaimed and subject to special police provisions. Disturbances spread to the neighbouring Bath estate in 1851, coinciding with the arrival of a new agent (Trench once more), who embarked on a programme of ejectments, assisted emigration and clearance of rent arrears. A handbill was posted, according to Trench, on every Catholic chapel on the estate:

> To Landlords, Agents, Bailiffs, Grippers, process-servers, and usurpers or underminers who wish to step into the evicted tenants property, and to all others concerned in Tyranny and Oppression of the poor on the Bath Estate
> TAKE NOTICE
> That you are hereby (under pain of a certain punishment which will inevitably occur) prohibited from evicting tenants, executing decrees, serving process, distraining for rent, or going into another's land, or to assist any tyrant, Landlord, Agent in his insatiable desire for depopulation.... (Trench, 1868, p. 126)[30]

Many of the Bath estate officials were in constant danger during the 1850s from unknown elements in the tenantry. Trench and his son usually moved about the property (as they also did in their land agency in the King's County) with 'a brace of pistols' and a police patrol; Trench was convinced of what he called a conspiracy of Ribbonmen determined to assassinate him. His *Realities of Irish Life* contains a reconstruction of his mock trial by disaffected inhabitants of the estate.

By the late 1840s the tenantry were beginning to resort to democracy in their campaign of resistance throughout the country. Lords Palmer and Londonderry issued eviction notices to tenants who had refused to vote in the 1852 election in accordance with the instructions of the estate administration (Byrne, 1996, pp. 57-59). Poor Law guardians were elected to the workhouse by rate paying tenants who increasingly ignored the estates' directions to support their candidates. The Shirley correspondence from 1850 is preoccupied with this shift in the balance of power, E.J. Shirley fulminating in August about the election of Guardians 'who can hardly write their names and who cannot read'. In January he complained indignantly about the election of a national schoolmaster as Master of the workhouse. His agent George Morant was there 'and exclaimed against this election but of course that was of no use ... as he was the only dissentient. Kennedy of course [the Bath agent until late 1850] voted with the mob, no gentleman but George was present, he now is about to write to the Commissioners to ... try to get them not to sanction the election'.[31] The hostile *Dundalk Democrat* summarized the changing balance of power, noting that Shirley,

> had no power in the workhouse of Carrickmacross, for you have deprived him of all authority. His agent, when he storms and threatens in it, is only laughed at; and it is but a few weeks since a vote of censure was passed on him for his unmannerly conduct and English insolence to those guardians whom you have returned to protect your interests.[32]

A petitioning system, introduced during Trench's agency on the Shirley estate in 1843, represents an important formal encounter between tenant and landlord and illustrates many of the elements of subaltern relations with a dominant elite. Apart from its value as a record of the extreme poverty of many of the tenantry in the

1840s, the petitions also demonstrate a degree of naïve ingenuity on the part of tenants in the context of the rigour of estate regulations. One tenant sought permission in March 1845 to build a house:

> As your petitioner is now cast on the benevolence of the world without a Cabin to shelter him his cousin with your honour's permission offers him a spot to build on. That your petitioner humbly hopes that your honour will take his distressed state into consideration and graciously please to grant permission.[33]

Most of the petitions depict universal attitudes of subordination, particularly in the frequency of 'your honour' as a form of servile address, though in some cases one suspects a mask of subservience in the certain knowledge of the espionage system which kept the estate informed of their circumstances. Palmer's recalcitrant tenants petitioned against their threatened eviction in 1852, 'humbly hoping' that 'your honour will still continue them as tenants' (Byren, 1996, pp. 56-57). Ruth-Ann Harris has suggested that many of the women on the estate used the petitions as a means of subverting the patriarchal authority of fathers, brothers and sons by appealing to estate regulations when it suited their circumstances (Harris, 2000).

Satirical songs and ballads were part of a long tradition in Ireland and many were employed as vehicles of resistance, indicting and lampooning the landlord class. At the annual dinner in Shirley's mansion in October 1850, some of the tenants sang in Irish for the assembly: the *Dundalk Democrat*, which was hostile to Shirley and Bath, suggested that they were abusing Shirley and his officials to their faces (Broehl, 1965, p. 68). Numerous ballads and humorous verses marked popular hostility to the Trenches on the Lansdowne estate in 1858 (Lyne, 2001, pp. 297-298). Estate 'marriage laws' were particularly satirized in song, as for instance on the Bath estate in the 1860s:

> O girls of Farney is it true/That each true-hearted wench
> Before she weds must get consent /From pious Father Trench?
> O search green Erin through and through/And tell me would you find
> Match-maker and land agent too/In one small farm combined?[34]

Overarching all of these strategies of resistance universally is language and what has been significantly characterized in colonial situations as 'native cunning', both fundamental in the armoury of the powerless. This is well illustrated in a confusion of language and landscape which was associated with many landed estates and the relations between tenant and colonial elite. In describing the densely-populated rundale landscapes of the west of Ireland, Estyn Evans suggested that the 'word used to describe the confusion of innumerable scattered plots and tortuous access ways … was 'throughother' [*trína chéile* in Irish], a word which has often been applied to other aspects of Irish life'. This confusion of landscape was exacerbated by nicknames for numerous families sharing surnames in the locality. Most landowners and agents saw 'the townland and their settlements as merely another obstinate obstacle to any rational management … an occult device that muddled responsibilities between master and tenant, perpetuated the old listless ways, and bred conspiracy' ... 'Its very incoherence was their protection ... Their means of resistance – conspiracy, pretence, foot-dragging, and obfuscation – were the only ones available to them, 'weapons of the weak' like those employed by defeated and colonized people everywhere' (Evans, 1973, p. 60; see also Scally, 1995, pp. 12-13). These intricately occupied, named and fragmented landscapes containing elaborate federations of kin groups confused many colonial prescriptions for neatness, order and civilization, especially in situations where there was no resident landlord.

Language as the expression of cunning, especially if not the language of the colonist, was a powerful weapon, masking true intentions and feeding into the 'oriental inscrutability' of colonial discourse. Hiberno-English, and its comic or 'charming' convolutions in 'blarney,' for example, may be seen as having a subtle objective of obfuscating or subverting the language of the colonial elite. 'Lying to the landlord' is part of local folklore in many parts of the country, playing on the understanding that the landowning elite assume the same rules of 'civilized' behaviour apply to all (McLoughlin, 1999, p. 192). Mimicry, mockery and satire also fall into the same strategic use of language as a weapon of resistance. However, as estate correspondence and official colonial archives elsewhere show, many of the dominant elite were aware of such tactics. Trench knew that the tenants on the various estates he managed were, as he suggested, 'servile and fawning whilst under restraint'. As he informed Lady Bath in 1851, he had no intentions of being 'humbugged' by outward signs of welcome.

'Native cunning' is perhaps the best manifestation of the 'Other' in colonial situations, a term which has been transmitted through estate records and formal narratives of the Irish landed elite, deeply signifying many of the elements and dispositions of colonial discourse. With the authority of the [extensive] written record of the landed gentry preserved today in public and private archives, the mindset behind 'cunning' still echoes down to us as a righteous term demonstrating the rectitude of the rulers and the intransigence and deviousness of the ruled. Some researchers may unquestioningly or unconsciously appropriate the outlook of the dominant class who wrote the narrative and shaped the story. Alternative perspectives, aimed at restoring the integrity of the original insubordinate presence however, are difficult to access. There are more than a dozen Patrick Duffys in the 19th century records of the Shirley, Bath and neighbouring estates, whose voices are silent. A critical reading of the records suggests that for these weak and absent witnesses, innate cunning was an important part of an armoury of resistance and survival in a world where the chips were generally stacked against them.

It may be argued that the system of landed estates was a fundamental component of the colonial project in Ireland from the 16th to the 19th centuries. Its owners and managers aimed for order and progress, articulated generally as 'improvement' and 'civilization' in estate landscape and society in the 18th and 19th centuries. Many of the struggles in 18th but especially 19th century rural Ireland can be seen as a collision between the top down intentions of the powerful landowning elite for neat and ordered landscapes, neat and docile tenantry, regulation, control and asset management, and local, impoverished and untidy tenant resistance. Resistance ranged from anonymous conspiracy, assaults and murder, to a more general practice of feigned subservience and obfuscation, often characterized by the landowning class as fecklessness, wiliness or 'native cunning' – the ultimate weapon of the powerless.

Notes

1 On what is still referred to by many in Northern Ireland today as 'the mainland'.
2 Public Research Office of Northern Ireland (hereafter PRONI) Belfast, D3531/C/2/2 Shirley papers. Mrs Shirley (Ettington Park) to Evelyn Philip Shirley, 20 April 1856. Papers quoted by kind permission of Major Shirley, Lough Fea.
3 PRONI, C/3/1/4 Frances Shirley to E.J. Shirley, 20 February 1832.
4 PRONI, D3531/C/2/1. E.J. Shirley to E.P. Shirley, February 1848.
5 NLI Ms 1420 quoted in A. Doyle (2001, p. 37).
6 PRONI, D3531 C/3/1/6. Smith to Mitchell, Oct 1839.
7 Thomas Carlyle, *Reminiscences of my Irish journey* (London, 1882) quoted in John Crowley (2003). I am grateful to John Crowley for permission to refer to his doctoral thesis.
8 PRONI, D3531/C/3/5. Sudden to Trench, 5 May 1843.
9 Longleat Library. Bath papers, Irish Box iii. Shirley to Lord Wilburton, 19 September 1850. Quoted by kind permission of the Marquess of Bath.
10 Longleat Library. Irish Box iii. Trench to Lady Bath (encl.), 21 December 1850.
11 Longleat Library. Irish Box iii. Trench to Lady Bath, January 1851.
12 Longleat Library. Irish Box iii. Enclosed with Trench to Lady Bath, 26 February 1851.
13 Estate agent Barbara Verschoyle to Lord Fitzwilliam, 1801, from W. Nolan and A. Simms (eds), (1998, p. 132); *Greig's Report on Gosford Estates*, p. 104.
14 PRONI, D3531/C/3/1/7, Shirley draft address, November 1839.
15 Handbill, November 1838, quoted in Broehl (1965), p. 44.
16 Kenmare estate minute book, quoted in G. O'Connor, The Lansdowne estate, 1848-58: the Poor Law, emigration and estate management, unpublished MA thesis, University College Dublin, 1994, p. 121.
17 PRONI, D3531/B/1. Farney Bubble Book, printed handbill September 1842.
18 Longleat Library, Irish Box iii. Kennedy to Lady Bath, 5 July 1849.
19 PRONI, D3531/S/55 Shirley papers, Trench's report, reproduced in Duffy.
20 Vaughan, *Landlords and Tenants*, p. 104; Irish Folklore Department, Schools Collection, Magheracloone parish schools.
21 PRONI, D3531/A/4. Observations on future letting of estate by Norman Steel, November 1795.
22 PRONI, D3531/C/2/1. EJS to EPS, 1 September 1850.
23 Longleat Irish papers, Bath estate annual report, 1853.
24 PRONI, D3531/C/3/1/7. Draft address of E.J. Shirley, 14 November 1839.
25 PRONI, D3531/C/3/5. Gibson, 8 May 1843.
26 Based on Trench, *Realities*; Broehl, *Molly Maguires*; and PRONI. Shirley papers, C/3/5.
27 From Farney Bubble Book, Shirley papers, cited in Broehl, *Molly Maguires*, p. 54.
28 PRONI, D3531/P/3, undated, [c.1844].
29 See Broehl, (1965), p. 66.
30 Trench, 1869, p. 126. As Kevin Kenny (1998, p. 15) has suggested the text of this notice was probably edited by Trench.

31 PRONI, D3531/C/2/1. See Lyne, *Lansdowne Estate*, 611-614 for similar trends in Kerry in 1858.
32 Longleat Library, Bath papers, printed notice from editor, *Dundalk Democrat*, September 1850.
33 PRONI. D3531/P/ box 1. Tenants petitions.
34 *Dundalk Democrat* 12 June 1869. A year after Trench's death in 1872, his ornate headstone was broken under cover of darkness. A recent restoration attempt was similarly destroyed.

References

Blacker, W. (1837), *Essay on Improvement to be Made in the Cultivation of Small Farms*, Royal Dublin Society, Dublin.

Bric, M.J. (1985), 'The Whiteboy Movement in Tipperary, 1760-80', in W. Nolan and T. McGrath (eds), *Tipperary: History and Society*, Geography Publications, Dublin, pp. 148-84.

Broehl, W.G. (1965), *The Molly Maguires,* Harvard University Press, Cambridge, Massachusetts.

Bowen, E. (1942), *Bowens Court and Seven Winters*, Cuala, New York.

Byrne, D. (1996), 'The Impact of the Great Famine on the Palmer Estates in Mayo', unpublished MA thesis, National University of Ireland, Maynooth.

Chatterjee, P. (2000), 'The Nation and its Peasants', in V. Chaturvedi (ed.), *Mapping Subaltern Studies and the Postcolonial*, Verso, London, pp. 8-23.

Clark, S. (1979), *Social Origins of the Irish Land War*, Princeton University Press, Princeton.

Clark, S. and Donnelly, J.S. (eds), (1983), *Irish Peasants: Violence and Political Unrest 1780-1914*, Manchester University Press, Manchester.

Crowley, J. (2003), Representing Ireland's Great Famine (1845-1852): A Cultural Geographic Perspective, unpublished PhD thesis, University College, Cork.

Day A. and McWilliams, P. (eds.), (1998), *Ordnance Survey Memoirs: Counties of South Ulster 1834-8,* Institute of Irish Studies in Assistance with Royal Irish Academy, Belfast.

Donnelly, J.S. (1975), *The Land and the People of 19th Century Cork: The Rural Economy and the Land Question*, Kegan Paul, London.

Dooley, T. (2001), *The Decline of the Big House in Ireland*, Wolfhound Press, Dublin.

Doyle, A. (2001), *Charles Powell Leslie II's Estates at Glaslough, County Monaghan 1800-1841*, Maynooth Studies in Irish Local History, Irish Academic Press, Dublin.

Duffy, P.J. (1997), 'Management Problems on a Large Estate in Mid-nineteenth Century Ireland: William Steuart Trench's Report on the Shirley Estate in 1843', *Clogher Record,* vol. xvi, pp. 96-100.

Duffy, P.J. (2004), 'The town of Monaghan: A Place Inscribed in Street and Square', in E. Conlon (ed.), *Later On: The Monaghan Bombing Memorial Anthology*, Brandon, Dingle, pp. 14-32.

Edgeworth, M. (1895), *Castle Rackrent and the Absentee*, MacMillan, London.

Evans, E.E. (1973), *The Personality of Ireland: Habitat, Heritage and History*, Cambridge University Press, Cambridge.

Foster, R.F. (1988), *Modern Ireland 1600-1972*, Lane, London.

Friel, P. (2000), *Frederick Trench and Heywood, Queen's County: The Creation of a Romantic Demesne*, Maynooth Studies in Irish Local History, Four Courts Press, Dublin.

Harris, R. (2000), 'Negotiating Patriarchy: Irish Women and the Landlord', in M. Cohen and N.J. Curtin (eds.) *Reclaiming Gender: Transgressive Identities in Modern Ireland*, St Martins Press, New York, pp. 207-26.

Hill, M. (2002), 'Investigative History: A Case Study – Lord Farnham and the 'Second Reformation', in B.S. Turner (ed.) *The Debateable Land: Ireland's Border Counties*, Ulster Local History Trust, Downpatrick, pp. 76-83.

Johnson, N. (1996), 'Where Geography and History Meet: Heritage Tourism and the Big House in Ireland', *Annals of the Association of American Geographers*, vol. 86, pp. 551-56.

Jupp, P. (1973), *British and Irish Elections 1784-1831*, David Charles, Newton Abbot.

Kenny, K. (1998), *Making Sense of the Molly Maguires*, Oxford University Press, New York.

Kiberd, D. (1995), *Inventing Ireland: The Literature of the Modern Nation*, Vintage, London.

Kiely, M.B. and Nolan, W. (1992), 'Politics, Land and Rural Conflict in County Waterford, c.1830-1845' in W. Nolan and T.P. Power (eds), *Waterford: History and Society*, Geography Publications, Dublin, pp. 459-94.

Leslie, S. (1923), *Doomsland*, Chatto and Windus, London.

Lester, A. (1998), *Colonial Discourse and the Colonisation of Queen Adelaide Province, South Africa*, Historical Geography Research Series No. 28, London.

Lyne, G. (2001), *The Lansdowne Estate in Kerry Under the Agency of William Steuart Trench 1849-72*, Geography Publications, Dublin.

Maguire W.A. (ed.) (1974), *Letters of a Great Irish Landlord: A Selection from the Estate Correspondence of the Third Marquess of Downshire, 1809-45*, Public Record Office of Northern Ireland, Belfast.

McGrath, T. (1992), 'Interdenominational Relations in Pre-famine Tipperary', in W. Nolan and T. McGrath (eds), *Tipperary: History and Society*, Geography Publications, Dublin, pp. 256-87.

McLoughlin, T. (1999), *Contesting Ireland: Irish Voices Against England in the Eighteenth Century*, Four Courts Press, Dublin.

Nolan, W. and Simms, A. (eds), (1998), *Irish Towns: A Guide to Sources*, Geography Publications, Dublin.

O'Connor, G. (1994), 'The Lansdowne Estate, 1848-58: the Poor Law, Emigration and Estate Management', unpublished MA thesis, University College Dublin.

O'Mearáin, L. (1981), 'Estate Agents in Farney: Trench and Mitchell', *Clogher Record*, vol. X, p. 409-12.

Proudfoot, L.J. (1993), 'Spatial Transformation and Social Agency: Property, Society and Improvement, c.1700 to c.1900', in B.J. Graham and L.J. Proudfoot (eds), *An Historical Geography of Ireland*, Geography Publications, London, pp. 219-57.

Proudfoot, L.J. (1997) 'Landownership and improvement ca.1700-1845' in L.J. Proudfoot (ed.) *Down: History and Society*, Geography Publications, Dublin, pp. 203-37.

Proudfoot, L.J. (2000), 'Hybrid space? Self and Other in Narratives of Landownership in 19th Century Ireland', *Journal of Historical Geography*, vol. 26, pp. 203-21.

Proudfoot, L.J. (2000), 'Place and *Mentalité*: the 'Big House' and its Locality in County Tyrone', in C. Dillon and H.A. Jefferies (eds), *Tyrone: History and Society*, Geography Publications, Dublin, pp. 511-42.

Proudfoot, L.J. (2001), 'Placing the Imaginary: Gosford Castle and the Gosford Estate, ca.1820-1900', in A.J. Hughes and W. Nolan (eds), *Armagh: History and Society*, Geography Publications, Dublin, pp. 881-916.

Routledge, P. (1997) 'A Spatiality of Resistances: Theory and Practice in Nepal's Revolution of 1990', in S. Pile and M. Keith (eds), *Geographies of Resistance*, Routledge, London, pp. 68-86.

Routledge, P. (1997a), 'Survival and Resistance' in P. Cloke, P. Crang and M. Goodwin (eds.) *Introducing Human Geographies*, Arnold, London, pp. 76-83.

Saris, A.S. (2000) 'Imagining Ireland in the Great Exhibition of 1853', in G. Hooper and L. Litvack (eds), *Ireland in the Nineteenth Century: Regional Identity*, Four Courts Press, Dublin, pp. 66-86.

Scally, R. (1995), *The End of Hidden Ireland: Rebellion, Famine and Emigration*, Oxford University Press, New York.

Scott, A.T. (2003), 'An Old Established Firm': Somerville and Ross and the Literary Expression of Social Change in Ireland, 1880-1925', unpublished MA thesis, Queens University Belfast.

Scott, J.C. (1976), *The Moral Economy of the Peasant: Rebellion and Subsistence in Southeast Asia*, Yale University Press New Haven.

Scott, J.C. (1990), *Domination and the Arts of Resistance*, Yale University Press, New Haven.

Somerville, E. and Ross, M. (1901), *Some Experiences of an Irish R.M.*, Longmans, Green and Co., London.

Somerville, E. and Ross, M. (1920), *Mount Music*, Longmans, Green and Co., London.

Somerville-Large, P. (1995) *The Irish Country House: A Social History*, Sinclair-Stevenson, London.

Swift, J. (1953), *Gulliver's Travels*, Collins, London.

Thompson, F.M.L. and Tierney, D. (1976), *General Report on the Gosford Estates in County Armagh 1821 by William Greig*, Public Records Office of Northern Ireland, Belfast.

Trant, K. (1997), 'The Landed Estate System in the Barony of Talbotstown Lower in the Nineteenth Century', unpublished MA thesis, National University of Ireland, Maynooth.

Trench, W.S. (1868), *Realities of Irish life*, Longmans, Green and Co., London.

Vaughan, W.A. (1994), *Landlords and Tenants in Mid Victorian Ireland*, Oxford University Press, Oxford.

Whelan, K. (2004), 'Reading the Ruins: The Presence of Absence in the Irish Landscape', in H.B. Clarke, J. Prunty and M. Hennessy (eds), *Surveying Ireland's Past: Multidisciplinary Essays in Honour of Anngret Simms*, Geography Publications, Dublin, pp. 297-328.

Yeoh, B. (2000), 'Historical Geographies of the Colonised World' in B. Graham and C. Nash (eds), *Modern Historical Geographies*, Pearson, Harlow, pp. 146-66.

Chapter 3

The Unsettled Country: Landscape, History, and Memory in Australia's Wheatlands

Joy McCann

Introduction

> That is the true test of a vital culture – to be able to sift through earlier achievements and rediscover new ways of seeing it, or us, or the world we live in, this 'place' that we all take as a map for our journeys (Shapcott, 1991, p. 23).

A new century dawns. Another drought unfolds in eastern Australia. This drought is shaping up to be as bad as they get. Comparisons are made with the last big 'dry' in 1990, and the ones before that in living memory. One hundred years ago, the inland regions of Australia were in the terminal stages of a drought that lasted, with little relief, for six years. Drought and death are part of life in the semi-arid landscapes of the Australian wheatlands (see Figure 3.1).[1] As dry conditions prevail, weather maps are scrutinized, crops germinate and falter in the paddocks, and stock and kangaroos compete for dwindling grasses. Dust storms whip ancient soils into life. The whole landscape seems to be on the move as westerly winds drive the lighter particles eastward, towards the densely populated coastal fringes. Perhaps city dwellers will taste the drought and contemplate the physical realities of the inland. Perhaps the dust storms are also symbolic of a more general exodus, as people vacate their farms and towns and drift coastward, or resettle in regional centres offering greater social and economic opportunity. They may reflect on the fact that soil and population drift are simply a fact of life in the settled landscapes of the inland.

Australia is one of the most highly urbanized nations in this world, yet rural mythologies continue to shape contemporary discourses about national character and heritage. This chapter addresses two particularly potent ecological and cultural mythologies embedded in the post-colonial wheatlands. Whilst they may have equipped settler Australians with purpose and vitality in the colonizing process, these same constructions have also provided a powerful moral justification for the massive environmental damage and processes of indigenous dispossession instituted during the 20th century.

Australia's wheat-growing country forms an elongated crescent that brackets the vast inland of the continent from Western Australia to southern Queensland. In this region of slopes and plains, low hills rise suddenly then fold away into an indefinite horizon. The wheatlands region is generally regarded as an economic zone, dedicated to the production of agricultural and pastoral commodities for world markets. It is also perhaps the region most profoundly affected by changing global economic circumstances in the late 20th century. These cropped and grazed landscapes are neither mountainous, nor coastal, nor desert. Australian geographer J.M. Powell (1988, p. 264), calls them Australia's 'landscapes of hopes' or, more bluntly, 'our monuments to an economic impertinence, sometimes confirming our fretful estrangement from Nature'. 'Perhaps the most salutary lessons of the postwar era', he wrote in a history of the Australian and New Zealand Association for the Advancement of Science published during Australia's Bicentennial year, 'are encountered in the intermediate spaces between the urban and outback or interior zones of Australia, and in the established farming districts of New Zealand'.

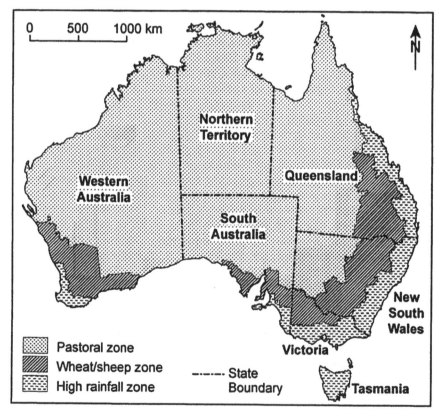

Figure 3.1 Australia's Wheatlands, or Wheat/Sheep Zone

These intermediate spaces loom large in the psyche of urban Australia. Many older Australians were nurtured on school and media images depicting a rural prosperity and abundance that flowed out to the coastal cities from the vast, swaying paddocks of the inland. In the 1950s and 1960s the crescent of wheat and sheep farming, sparsely-settled and infernally dry, was imagined as a tide of progressive agricultural and pastoral enterprize in an acquiescent, if not benign, landscape. This was an unruly country, progressively brought to heel by 20th century technology and scientific know-how. The agricultural and pastoral landscapes represented the public face of the nation, portraying its true character and spirit. In contemporary Australia, where two thirds of the population now live in cities and towns near the coastline, the rural landscape remains deeply inscribed with a colonial narrative about the heroic occupation of a difficult environment. This 'national rural myth' memorializes the story of settlers who arrived in the first years of the colony and later became selectors of rural land from which they struggled to make a living (Hirst, 1978; 1982). The 'pioneer' story acquired its particular mythological potency during the years of the late 19th and early 20th centuries, when the process of federating the colonies set the scene for a flourish of nationalistic story-telling about colonial conquest.

A related and equally robust rural myth also emerged at this time, celebrating qualities of mateship and egalitarianism attributed to itinerant male workers engaged in the Australian pastoral industry. The 'bush' legend was first articulated and popularized through the pages of the *Bulletin* journal – most notably by A.B. (Banjo) Paterson and Henry Lawson – and later revived by historians such as Vance Palmer in *The Legend of the Nineties* (1954) and Russel Ward in *The Australian Legend* (1958). They argued that the settlers who toiled in the remote areas of the inland embodied the very essence of an Australian national identity. Caught up with the optimism of post-war reconstruction, they tapped into a renewed sense of national unity in the face of adversity. They were, as Walter puts it, 'impelled by a wish to overcome the despair and conflict of the 1930s', envisioning a new social order in the rhetoric of the 'bush' legend (Walter, 1990, p. 78).[2] Kapferer (1990, p. 91) observes that the image of the 'bush' as the epitome of Australia, and of Australians as embodying mythological virtues associated with the colonized rural landscape, 'remains as fundamental to our understandings and mystifications of ourselves, in relation to each other and to outsiders, as it has ever been'.

In the 1950s, the agricultural landscape was promoted as a place with no past. In the post-war climate, it offered a blank slate primed for a new phase of closer settlement, and equipped with the new tools of technology and science from which the potential fruits of reconstruction would be reaped. If the difficulties and failures of the 1930s and 1940s were etched into the memories of farmers, the solutions offered by technology and science empowered them to reinvent the heroic pioneering narrative as their own. For most Australians, the history of colonization was neatly contained within a consensus of mainstream political and social narratives. In the wheatlands the complexities of environmental change, and the experiences of marginalized groups, barely made a ripple on the expansive oceans of grain. Manning Clark's *History of Australia* series, begun in 1962, foreshadowed a dramatic shift in Australian history-writing. By 1988, as Australia celebrated the Bi-centennial of European occupation, the 'triumphalist march of history' (Fitzgerald, 1990) was being radically rewritten by historians focusing on the impact of colonization on people and land. They opened a floodgate to histories about local places and the 'wilful obscuring of the experience of most people' since colonization (Walter, 1990, p. 80).

By the end of the century, rural Australia was in danger of overflowing with history, although the undercurrent of old mythologies persisted alongside the new narratives. The Australian heritage movement,[3] in particular, played a powerful role in keeping them alive, aided and abetted by a proliferation of community histories, tourist literature, and local history museums. The stories of settler occupation were writ large in the physical evidence of this transformation. Farms, woolsheds, cemeteries, and the enterprizes of town-building were lauded as evidence of a recalcitrant landscape transformed into the nation's productive heartland (Ireland, 1995, pp. 92-3). Even as the landscape repeatedly failed to measure up to expectations of easy abundance during the course of the century, the 'mythological solution' in the struggle to subdue and settle a recalcitrant environment was to recast the colonizing enterprize as a spiritual triumph (Gibson, 1992, p. 88). In this same period, however, the statistics indicated a waning in the economic and political influence of the nation's agricultural and pastoral sector. In 1950-51, for example, rural exports (primarily from agricultural and pastoral commodities) made up 85.5 per cent of Australia's total export market. By 1991-92 this had fallen to 23.1 per cent (Commonwealth of Australia, 1996, p. 8).

But it is the emptying of the rural landscape that has come to represent the face of rural decline. Census records, particularly after World War II, painted a graphic portrait of a steady, inexorable erosion of farm and small town populations, employment opportunities, and social infrastructure in inland Australia. The Australian Bureau of Agricultural Research and Economics (ABARE) calculated that the number of farming enterprizes declined by an average of 3.5 per cent per annum over the second half of the century, from 205,000 in 1955 to 125,000 in 1995 (Australian Bureau of Statistics, 1992).[4] In the mid-1950s, farm work constituted 15 per cent of the national workforce in Australia. By the mid-1990s that figure was reduced to just 5.1 per cent. The total population of people on farms throughout Australia fell by about 32,000 in this period (Australian Bureau of Statistics, 1993). Whilst Merrett (1977) demonstrated that the centripetal flow of people from rural to urban areas has been a relatively constant phenomenon since the mid-19th century goldrushes in Australia, the post-war statistics blandly catalogued the enormity of rural demographic change experienced within the lifetime of just one generation. In 1999, the Federal Government conducted an inquiry into the impact of its National Competition Policy. The inquiry revealed the extent to which smaller towns were 'withering' across the inland (Productivity Commission, 1999). The 'dying town' syndrome was most dramatic and visible in the wheat belt and dryland grazing regions exposed to international agricultural commodity markets. The regions most seriously affected were concentrated in the Mallee and Wimmera districts of western Victoria, the wheat-growing districts in the central and western regions of New South Wales, and in south-western Western

Australia. Meanwhile, larger regional centres, so-called 'sponge' cities, were soaking up people relocating from the scattered farming settlements, strangling the life of small towns and villages (McKenzie, 1994).

'Rural decline' remains a powerful metaphor for the precarious nature of settler habitation outside the metropolis in Australia. The discourse is closely linked to the financial status of farm businesses and the productivity of the agricultural sector. It gains momentum from the 'hard' evidence contained in statistical research conducted by the Australian Bureau of Statistics and the Australian Bureau of Agricultural Research and Economics. Rural decline is defined predominantly in terms of an industrialized model of agriculture, where the fate of rural communities is fundamentally tied to the economic performance of rural producers. As rural sociologist Stewart Lockie observes, '[v]irtually unquestioned is the assumption that agriculture is first and foremost an economic activity' (Lockie, 2000, p. 23). Rural policymaking and research continue to regard the economic development, social status, and technological creativity of rural producers as their core business (for example, see Reeves, 1990). Meanwhile, the people of the wheat belt are caught up in a discourse of rural decline, in which ecological degradation, social fragmentation, and Indigenous land claims profoundly unsettle conventional ways of seeing the inland. Whilst images of a rural arcadia continue to inspire nationalistic projects about the inland,[5] rural communities, struggling with a restructured domestic agricultural sector and declining terms of trade, seek to reinvent themselves as bastions of the pioneer legend.

These historical narratives, as deeply embedded as they are in rural settler mythologies, are fraught with contradiction and tension. Prolonged drought dramatically exposes the ecological and social stresses of farming landscapes and communities at the frontiers of rural life in Australia (Ramson, 1998, p. 174). Fault-lines criss-cross the wheat belt, creating ideological rifts and tremors in the national psyche. Competing images of rural abundance and decline jostle in the public imagination. Farming settlements that sprouted promisingly in the early 20th century to form the hub of a thriving hinterland were, by the end of it, more likely to host a cluster of derelict commercial and civic buildings. Within just one generation, farming families that enjoyed the financial and social gains of an economic boom in agricultural and pastoral commodities in the years after World War II found themselves in a rapidly transformed economic and social landscape. This was no longer fertile ground for a simple narrative espousing masculine resilience and egalitarianism in the face of a harsh and unforgiving landscape.

One of the most significant challenges to the prevailing economic assessment of the rural landscape is posed by mounting scientific evidence of ecological change in agricultural and pastoral regions. Criticisms of the effects of European land use practices on the quality of soils and vegetation were in circulation from the 19th century, and debates about whether they improved or degraded the environment continued, largely between the pages of journals and newspapers, throughout the 20th century. The debates were at their most vitriolic when economic downturn in agricultural trade coincided with drought conditions. However, from the 1970s, environmental scientists began to focus on ecological degradation to construct compelling arguments about land use and landscape change. The rise of ecological consciousness in Australia (Robin, 1998) coincided with a period in which broadacre agriculture was becoming increasingly expansive. Public discontent with the physical manifestations of degraded agricultural and pastoral landscapes escalated amid increasingly politicized environmental debates. In the 40 years between 1955 and 1995, the total area of land being cropped grew by an average of 6.4 per cent per annum, with the greatest increases occurring from the 1960s (Commonwealth of Australia, 1996).[6] By 1996, agriculture was the dominant form of land use, occupying 60.9 per cent of Australia's land surface (Commonwealth of Australia, 2001). Farmers were under financial pressure to increase their productivity, and this demanded the acquisition of more land for crops, and the adoption of new agricultural products and technologies to maximize yields.

Critics of agricultural practices focused on the scale and techniques of modern farming as a significant factor in the degradation of arable land and inland rivers. By the end of the 20th century, there was a growing consensus amongst environmental commentators that the inland agricultural and pastoral regions were experiencing environmental problems on an unprecedented scale. Indeed, the Commonwealth Scientific and Industrial Research Organization predicted that half a million hectares of the wheat belt in the eastern states alone had less than 50 years of productive life remaining (Mercer, 1995, p. 251). In constructing the rural landscape as an ecosystem in crisis, the ecological sciences emerged as an influential voice in debates about the future of farming and the sustainability of agricultural and pastoral communities. David Carter (1994, 9) suggests that the environmental discourse has kept rural Australia at the centre of national consciousness. 'This discourse expresses a powerful, ambiguous myth of non-Aboriginal belonging to the land'. In this context, there is a compelling need to understand the cultural and historical dimensions of an industrialized agricultural landscape increasingly characterized in terms of impoverished soils, denuded paddocks, and degraded river systems.

Ecological Narratives

The Lachlan Valley lies in the central-west of New South Wales (see Figure 3.2) and straddles the wheat belt. Since European colonization, much of the drama of the relationship between settler and environment has been played out at the fringes of arable and grazing landscapes such as this. Noel MacArthur's grandfather travelled from northern Victoria in 1890 to take up his selection of a 3,000 acre block in the central Lachlan, just north of Condobolin (Noel MacArthur (pseudonym), pers. comm., 2002). The property, called 'Mogo', is still owned by the MacArthurs. It was typical of blocks subdivided from former grazing runs, having been extensively ring-barked by gangs of Chinese labourers in the previous decade to encourage the native grasses to proliferate. By the time the MacArthurs arrived, the landscape was a forest of skeleton trees, and scrub proliferated instead.[7] Whilst most selectors at this time grazed sheep and cattle, he favoured wheat-growing, and another round of ring-barking began. Over the next 60 years, the property was again cleared of vegetation and he cropped with horse teams until acquiring his first tractor in about 1939. By then 'Mogandale' was a typical wheat farm of the Depression era, overstocked and 'pretty well flogged'.

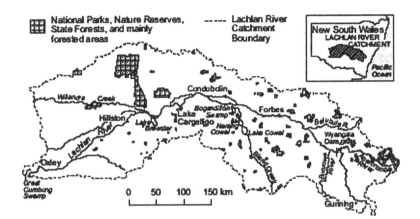

Figure 3.2 The Lachlan River Catchment

Fire was unwelcome in the fenced and stocked colonial landscape, and the propensity for particular species to regenerate in dense concentrations after tree clearing effectively reduced the amount of grass available for grazing stock and impeded cultivation. The explorer, Major Thomas Mitchell (1848, p. 413), had first observed the transformation of open forest lands to 'scrub' near Sydney in 1848. He noted how the suspension of traditional Indigenous burning practices by settlers allowed the growth of dense forest and grasslands 'choked by underwood'.

The negative effects of land clearing emerged as a matter of public debate in Australia from as early as the 1870s. In the early years of the 20th century, government policies encouraged and rewarded land clearing in order to ensure the development of agriculture, aided by scientific and technical advances and water conservation measures. Nevertheless, concerns about unfettered tree clearing were never far beneath the surface of public discourses about farming during the 20th century (Bonyhady, 2000, p. 85). In his classic study of the Pilliga Forest in northern New South Wales, Eric Rolls (1981) suggested that the grasslands and open woodlands observed by early colonists in eastern Australia were the result of traditional Indigenous burning practices, and that the denser forests were created as a result of European settlement. His thesis was later 'borrowed' by proponents of vegetation clearance on farmland, although Rolls himself argued in favour of valuing and conserving these 'phoenix forests' rather than removing them (Rolls, 1981, p. 399; Griffiths, 2002, pp. 382-3).

In January 1941, a series of articles by Arthur E. Heath appeared in *The Land*, highlighting the depletion of grasses and the proliferation of cypress pine

(*Callistria verrucosa*) following tree clearing across vast areas of farmland. While war raged in Europe and the Pacific, he wrote about Australians on the home front waging war with their own countryside. Native flora and fauna were widely condemned as 'Public Enemy No 1', and many of the closely settled districts were reduced to 'barren, treeless wastes' by indiscriminate use of the axe and firestick. Such criticisms, however, failed to sway the general predisposition towards widespread vegetation clearance in the emerging wheat belt. In the central Lachlan Valley, settler descendents still recall their loathing for the Mallee trees (*Eucalyptus socialis* and *E. dumosa*) that once knitted across the sandy plains. Their stories are peppered with the mixed emotions that accompanied the 'heroic pioneering labour of clearing' (Griffiths, 2002, p. 379). One farmer remembers his father clearing Mallee trees on his selection north of the Lachlan River, but is now uncertain about that legacy. 'I sometimes sit up the paddock, and look back to where we've cleared the country, and you can hardly remember what it was like before. Everyone that comes past says "gee, you've done a good job with that". [But now] I don't know what to think' (RS, pers.comm., 2001).

The tension between a desire to conserve the continent's natural resources and the pressure to develop and exploit them are persistent themes in Australia's ecological narratives since the earliest European encounters. Native vegetation legislation has merely served to crystallize the century-long conflict over the value of Australian indigenous flora and fauna, as public agencies and local communities wrestle with shifting and contested evaluations of the wheat belt. Depending on the values attributed to indigenous vegetation, extensive land clearing for grazing and cropping during the 20th century has either been extolled as evidence of scientific and technological prowess over the Australian environment, or condemned for causing degradation of the soils and waterways.

The continuing expansion of cropping into the inland plains since the 1950s represented perhaps the most dramatic episode of landscape change since European occupation. Ecological surveys of the wheat belt became increasingly common after the 1970s, and they signalled alarming rates of land and water degradation. In south western Western Australia for example, vegetation clearing progressed at a frenetic pace, and critics predicted that much of the land would be useless for agricultural purposes within 30 years as a result of increased salinity in the sandy soils. In 1990, the Chairman of the Australian Wheat Board spelt out the financial implications in *The Age* newspaper (9 May): Australia's wheat export market worth $A2 billion was in jeopardy as a result of reduced protein levels in wheat grown on nitrogen depleted soils.

Between the 1970s and 1990s, no less than 16 official inquiries called for a national policy on land use to address the now highly-charged political issue of land degradation resulting from two centuries of European farming and grazing practices (Horton, 2000, p. 13). Whilst it had long been 'common wisdom' that most native vegetation clearing occurred in the late 19th and early 20th centuries, these surveys suggested that just as much clearing was undertaken in the half century since World War II as in the previous 150 years (Commonwealth of Australia 1995, p. 1). Subsequent State government environmental reports condemned poor land management practices in agricultural regions: excessive clearing for cultivation, resulting in the fragmentation of vegetation cover, and neglect of remaining stands of trees and shrubs.

In the mid-1990s, the New South Wales National Parks and Wildlife Service published the results of a project to map native vegetation remaining in the State's wheat belt and to plot the rate of land clearing over a 15-year period. It revealed that, in the central Lachlan Valley, more than half of the native vegetation that existed in 1974 had been removed by 1989. While tree logging operations in the rainforests of South East Asia, the Pacific, and South America, and in southern Australia's temperate old growth forests were attracting widespread criticism, the fate of native vegetation in the local wheat belt landscapes had simply 'not been in the public eye' (Sivertsen, 1994, pp. 5-8). On current trends, native vegetation in the wheat belt would be 'negligible' by the end of the century. Land clearing is deeply entrenched in the agrarian culture of the wheat belt. 'At the heart of this issue is the notion that the Australian bush is inferior and unproductive' (Nadolny, *et al.*, 1995, p. 34).

It was in this context that the New South Wales government introduced the first of a suite of regulations in 1996 to control clearing of native vegetation particularly on privately-managed farmland. The State Environmental Planning Control 46 (generally referred to as SEPC 46), created enormous resentment amongst farmers opposed to the government controls on their farming activities especially at a time of increasing market pressure. However, government agencies implementing vegetation clearing controls confronted the prevailing belief that indigenous vegetation was not an issue of concern because it was previously cleared by earlier generations, and what remained was merely scrub. As intensive cropping gave way to a mixture of grazing and cereal farming, the pine trees returned with a vengeance. According to Noel MacArthur, the dense proliferation of cypress pine at the expense of other tree species 'really spoilt the balance' of the landscape. By the 1990s, the 'scrub terror' was once again at the centre of a political battle over vegetation management in the agricultural landscape.

With the introduction of the new State Native Vegetation Conservation Act 1997, the farmers of the central Lachlan found themselves at the front line of the native vegetation debate. The central wheat belt of New South Wales, with its recent history of accelerated land clearing, was selected for the first Vegetation Management Plan under the new State legislation. The plan aimed to improve and enhance native vegetation, and relied heavily on scientific evidence that the rate of vegetation clearing was exacerbating problems such as soil erosion, salting, and diminished indigenous flora and fauna on agricultural land. The legislative definition of native vegetation was all-encompassing, including trees, understorey plants, groundcover and wetland vegetation, comprising species 'that existed in the State before European settlement' (1997, Section 6). The Mid-Lachlan Native Vegetation Plan was predicated on the notion that 'remnant' vegetation ought to be conserved in order to replicate an earlier, and arguably more desirable, state at the time of European occupation. Conserving remnant native vegetation, then, required the committee to 'unearth a lot of history', as MacArthur puts it, in order to determine what constituted pre-European vegetation. He acquainted himself with the testimonies of explorers such as Mitchell and Evans, records of early grazing runs such as the Overflow Station north of the Lachlan,[8] and authoritative published accounts of the inland country including C.E.W. Bean's *On the Wool Track* (1945) written about the Darling River region. These sources enabled him to challenge what he saw as 'stereotypical concepts of what things might have been like when Captain Cook got here', and convinced him that Australia was more of a grassland than a forest land prior to settlement.

The wheat belt had become a battleground of competing ecological narratives, in which differing or selective interpretations of the past were promoted in order to frame a story that was coherent to the world view of its proponents. The Department of Land and Water promoted the new native vegetation plan by articulating it as a vision for the wheat belt landscape that would achieve an ecological 'balance' of native vegetation species with cultivation and grazing. The agency was keen to diffuse fears amongst farmers that the native vegetation planning process was the dead hand of government bureaucracy putting unnecessary and unworkable constraints on their livelihood. They argued that the fate of the central wheat belt's agricultural industries was ultimately linked to the fate of native vegetation on farmland, a point upon which those involved in the vegetation debate were deeply divided. It became known amongst local farmers in the central Lachlan Valley as the '1770' policy. The government's objective to restore elements of a pre-European landscape confronted and destabilized the settler narrative, whereby tree clearing was not just a physical necessity but a moral imperative to occupy the Indigenous landscape. According to Noel MacArthur's reading of historical sources, the combined forces of Aboriginal fires and lightning strikes would have created a mosaic of burnt and unburnt patches, maintaining the diversity of plant species and overcoming problems of uncontrollable wildfires. Such revelations convinced Noel that farmers had a case to argue for continuing some active form of vegetation management.

In response to the farmers' interpretation of historical sources, the Native Vegetation Advisory Council presented its own historical trump card. In a public brochure promoting the conservation of native vegetation, the people of New South Wales were urged to value the physical and spiritual significance of 'the bush' as 'part of our national psyche' and identity (New South Wales Government, c.1999). Linking the potent cultural symbolism of the 'bush' with native vegetation conservation seemed to offer some hope of shifting entrenched beliefs. The Council also commissioned an analysis of the social and historical significance of native vegetation, emphasizing the role of cultural values in shaping settler and Indigenous people's interactions with the biophysical aspects of the landscape. Australians, according to the authors, needed a shared vision of the rural landscape, a definition of 'what we want the "bush" to look like' (Lambert and Elix, 2000, p. 23). They advocated recognizing local landscape knowledge and building local pride in native vegetation as a crucial part of that vision. However, they also acknowledged that, with few Australians now having any direct links with agriculture, such a national vision would have to accommodate the competing priorities of groups that did not necessarily share the experiences and values of farmers.

Government environmental agencies and individuals concerned with promoting the conservation of native vegetation were embroiled in deeply entrenched cultural narratives about improving the land. Stories about landscape change in the wheat belt are rooted in the belief that modern farming practices have improved the ecological health of the land, the proof of which is to be observed in the resilience and sophistication of agricultural production. Productivity is regularly cited as proof of ecological health. In cultural terms, what emerges is a more complex response to the vegetation issue than recent studies of farmers and natural resource management suggest. By drawing on historical interpretations, scientific research, and local mythologies about landscape change, farmers are seeking to recast themselves in a modernized version of the heroic settler narrative. It entails a story of how they inherited a degraded landscape brought about by rabbit plagues and overgrazing culminating in the 'dust bowl' days of the 1940s, and how they succeeded in restoring it to productive health. They perceive themselves as modern-day successors to a much older tradition of Aboriginal land management that shaped and maintained a lightly treed, grassy landscape that had so appealed to the aesthetic and economic interests of the early colonists. Land clearing controls go to the heart of tensions between city-based environmental agencies and farmers over notions of rurality and the economic and cultural worth of rural life and work.

Nature's restorative power is a recurring theme in local stories about landscape change in the wheat belt. It hints that human impacts on the landscape may be only temporary after all. It suggests that nature has its own dynamic. According to this view of nature, the passing of time obliterates even the catastrophes of human mismanagement. That nature and humans are separate entities is one of the most powerful cultural assumptions underpinning these perceptions of the wheat belt landscape. According to T.H. Bath, writing in *The Land Farm and Station Annual* (1949, p. 37), 'most sons of the soil think of Nature as something apart from themselves: Man with a capital "M"; nature with a small "n"'. Bath, a farmer in the West Australian wheat belt, was likely to have been familiar with the sentiments of North American writer and former expert on game management, Aldo Leopold in his book *The Sand County Almanac* (first published 1949, it later became essential reading for adherents to the emerging environmental movement in the 1970s and 1980s). Both had witnessed the diabolical results of combined drought, economic depression, and land clearing in the Great Plains of the United States and the settled Australian inland respectively during the 1930s and 1940s.

Bath found time in his retirement to reflect on the 'higher intrinsic pleasures of communion with nature on the farm'. He urged farmers to get 'in tune with nature', to experience:

> the intoxicating fragrance of damp earth from the first good shower after the drought of summer; the sight of birds flying through the outer foliage of trees to drench their plumage, in a like [sic] exuberant delight; the sense of well-being in watching the long furrows of red or brown earth heeling over from the stroke of the ploughshares; the morning and evening carols of the magpies from the treelots conserved by the wise farmer...these are only some of the exquisite joys reserved for a more select few (Bath, 1949, p. 38).

In the Western Australian wheat belt that was Bath's country, settler farmers were enthusiastically carving fields from grassy woodlands in the 1940s, a practice that continued until well into the 1970s. Bath's article highlighted one of the defining conundrums for farmers in the mid-20th century. On the one hand, progressive farming methods required increasingly extensive areas of land to be denuded of trees and shrubs in order to grow crops on a large scale. On the other hand widespread vegetation clearance, particularly in combination with drought and overstocking, yielded disastrous environmental and social consequences. Bath's vision for the wheat belt was of farmers working in harmony with nature, where the 'original inhabitants' (referring to animals, birds and insects) would continue to 'reign with native assurance' as long as belts of trees and shrubs were allowed to remain. Drawing on stories from Classical mythology, he warned of nature's powerful agency, predicting that retribution would come to those who transgress its laws, urging farmers to embrace nature rather than to assert themselves over it. His vision was of nature taking vengeance on those responsible for what he saw as indiscriminate vegetation removal through burning, overstocking and ring-barking. It carried a strong message to post-war readers of *The Land* journal, who were more accustomed to articles on rising agricultural markets and the moral virtues of farming life. 'Only the land-miner who used the firestick with undiscriminating fury, and erected structures of wood and iron in a dust-plagued vacuum is left to the companionship of ants and flies and the harmful bugs and grubs that raid his crops' (Bath, 1949, p. 38). Warnings that the health of humankind is ultimately reflected in the ecological health of the land continue to reverberate in the wheat belt. As one Lachlan Valley farmer observed, 'I've always said if you look after nature she'll look after you, and you cross nature she'll kick you, and she certainly will' (MD, pers.comm., 2002).

Contested Terrain

About half way along the central Lachlan valley a stone cairn stands, built in 1914 near the old Kiakatoo Station. In the fine grey soils of the Lachlan floodplain, at a place called Gobothery Hill west of Condobolin, the rubble stone memorial marks the place where the colonial Surveyor-General John Oxley's exploring party camped on the site of an Indigenous Wiradjuri man's grave in 1817. According to Oxley's journal, a tribal leader drowned in floodwaters was buried here, his grave marked in the traditional Wiradjuri way with carvings etched into two cypress pine trees that stood like sentinels at the gravesite (Black, 1941, pp. 13-14).[9] The cairn marks the frontier of settler and Indigenous relationships in the central western plains of New South Wales. It now bears mute testimony to settler occupation of Wiradjuri country during the early 19th century.

In these poorly-watered plains, the river was also favoured by 19th century European settlers seeking grazing land. The river systems of the Murray-Darling Basin that flow westward from the Great Dividing Range became conduits for European exploration and settlement, just as they had ferried European diseases inland in advance of the first exploration parties. Competition for river country precipitated a protracted period of Aboriginal resistance and survival, as white settlers expanded their dominion over traditional Wiradjuri land. By the mid-19th century, small bands of Wiradjuri people still maintained a largely traditional existence, but many were increasingly drawn into contact with settlers who had claimed extensive tracts of pastoral land fronting the inland river systems.

From 1861, settlers began dissecting the country into a network of farms under new colonial land laws. As the catastrophic effects of settler occupation unfolded amongst the Indigenous populations of inland New South Wales, the colonial government established the Aborigines Protection Board in 1883. The Board was charged with administering the 'Indigenous problem', and initial policies focused upon the marshalling of those considered to be 'full-blood', frail or ill, into government-run Aboriginal reserves. In Wiradjuri country, the Board gazetted a series of these reserves or missions, extending along the Lachlan River from Condobolin through to Cowra, and along the Murrumbidgee River. The creation of the Board coincided with what Read (1988) describes as 'a hundred years war', lasting at least until the Board was finally abolished in 1969. During the early years of the 20th century many Wiradjuri, dispossessed of traditional life and country, were forced to base themselves in official reserves in return for survival rations. Some formed smaller communities on grazing properties or farms, or established camps on the fringes of local townships.

On 8 August 1914, as a small group of men gathered around the Oxley cairn for the unveiling ceremony, the Minister for Labour, spoke of the 'truly Australian national spirit' that this site represented; 'It was the bounden duty of all Australians to search for and preserve the earliest links of the discovery and opening up of Australia' (*Condobolin News*, 8 August 1914). By this time, the Wiradjuri people in the valley had been decimated, and many settlers were echoing official opinions that they, like Indigenous people elsewhere in colonized Australia, were a dying race. According to a classic myth of Western colonialism, traditional Indigenous connections with land died out in the wake of European invasion and occupation, while the material remains of their presence were collected and closely examined in order to unravel universal wisdoms (Rose and Clarke, 1997, p. 25).

The site of the Oxley memorial is one of the few known publicly commemorated Wiradjuri sites in the region. It offers visitors a glimpse of a mysterious Indigenous past, transmuted into a site of European occupation. This overlay of European explorer campsite and Indigenous grave neatly underscores the prevailing mythology about the severing of Indigenous connections with their traditional country. The Oxley expedition of 'discovery' and 'pioneering' becomes the beginning of historical time. Dominated by narratives of agricultural transformation, the Kalar country is silent in the settler imagination. The Indigenous people of the Lachlan Valley are safely relegated to 'prehistory', whilst the Kalar landscape is reconstructed as an artifact in which the physical marginalization of their descendants to the edges of town contributes to a perception that they are erased from the wheat belt landscape. The relics of the dispossessed are stored in local museums and private collections, and any allusion to the authenticity of contemporary Wiradjuri as 'real' Aborigines, and any meaningful connections between them and their ancestral country, are vigorously denied.

Published accounts of the region's history move swiftly through early European accounts of Indigenous tribal groups along the Lachlan River as a preface to the main story of settler occupation. Where Indigenous experiences are recounted at all, they construct a narrative about their 'passing away', the precursor for the 'more important drama of white settlement' (Cronon, 1992, p. 1364). W.E.H. Stanner (1969, p. 25) called it 'the great Australian silence'. This 'strategic forgetting' has become a way of remembering the past that legitimizes and normalizes the processes of European colonization and renders Indigenous people invisible in the colonized landscape. 'The land and the indigenous peoples become merged, the former foregrounded, the latter denied a place in history at all' (Curthoys, 2003, pp. 191-2; also see Otto 1993, pp. 545-8; Griffiths 1996, pp. 107-8).

During the last two decades of the 20th century, a new wave of Aboriginal histories and two landmark legal decisions in the 1990s[10] challenged the prevailing notion that Indigenous connections with land had effectively been erased by white settlement. In the process, they offered a radically different construction of the rural landscape as a place of continuous occupation and belonging. Legal recognition of native title brought to the surface of public consciousness long-festering racial discourses about the relationship between Indigenous and settler Australians in the regions of inland Australia dominated by agricultural and pastoral interests, manifesting in emotionally-charged debates about the legitimacy of European colonization. In particular, it cast the whole notion of an Australian 'bush' identity in a new light. The inland could no longer be constructed in terms of a simple narrative espousing settler resilience and conquest in the face of a harsh, unforgiving, and empty landscape. It was also a place of Aboriginal survival, connectedness, and spirituality.

A Wiradjuri woman who has lived all her life within sight and smell of the Lachlan River recounts a story about her son running home from school one day to tell her that the children had been raking out some sand delivered to the school and found human remains from an old burial ground. Aboriginal burials often took place in soft sand hills along the rivers and creeks. Tracing the source of the sand to a nearby property, she looked down into the pit and saw the remains of her people, exposed by the sand digging operations.

> They had the skulls, all smashed with bullet holes, and there was hundreds, all around embedded, all around the walls....I sat there, and you know I went down on my knees, I cried for them, and this wind come. And it was just like all of them was howling at the same time (BA, pers.comm., 2002).[11]

She mentions a local landmark that looks like a dead mountain but, if you walked up and looked over the other side, you would find ten rocks at the bottom, 'ten rocks, where the ten different tribes met all the time. And they sat there, the elders, they sat there and they discussed' (BA, 2001). Whilst Western archaeologists have been preoccupied with collecting and preserving the physical artifacts of the past, older Wiradjuri harbour vivid stories of new relationships formed with this 'dead' country of their ancestors, and express a deep sense of connection in ways that belie settler perceptions. Dispersed family groups have sought to resurrect the sanctity of old places by attaching new stories to the severed thread. Their stories give emphasis to continuity rather than separation, living social and spiritual connections rather than antiquity (Griffiths, 1996, p. 100; Lowenthal, 1985, xvii, pp. 410). As Read (1988, p. 77) found in his research on Wiradjuri history, 'the present and the past were the same country'. In these landscapes of enclosure, a sense of fluid connection survives tenaciously amongst the orderly grid of paddocks.

By the end of the 20th century, the Lachlan Valley was a cradle of contradictions and tensions: a prosperous agricultural and pastoral zone, an ecological system in crisis, a frontier where Indigenous and settler Australians shared an entangled but scarcely acknowledged history, and a storied place embedded with a rich lode of collective and personal memories. The life stories, in particular, offer insights into how people reconstruct their past, make sense of changing ecological and social conditions, and define their sense of place. The cultural dimensions of the wheat belt speak most eloquently through the spoken word. Selected stories and observations help to tease out the meanings embedded in particular places, how the past is remembered and valued in the present. We tell stories to make sense of our lives, and to place our experiences in a wider context. 'Stripped of the story, we lose track of understanding itself' (Cronon, 1992, 1369). Meanwhile, the stories of place and belonging lie embedded in river banks, and buried beneath sand drifts along old fence lines. Rural Australia remains a powerfully mythologized place, its stories capturing and sustaining intimate connections between land and people. They are central to understanding the cultural history of contemporary rural Australia.

As another drought unsettles the landscape, the legacy of transforming the drier zones of the Australian continent into productive farming landscapes is brought into sharp relief once more. There is an urgent imperative in Australia to examine the tensions inherent in the post-colonial landscape, to grapple with the cultural and historical dimensions of land degradation and the inseparably entwined legacies of setter and Indigenous history in rural Australia. The bones of tension and contradiction that lie at the heart of the settler Australian relationship with the inland are exposed in the desiccated soils. The difficulties that settler Australians have encountered in inhabiting this country compel us to return repeatedly to questions about our cultural preoccupations and relationships with the environment. George Seddon (1976, p. 16; 1997, p. 71; 1998) suggests that Australians are still learning to apprehend their physical environment and their place in it imaginatively. Settler mythologies associated with the triumphant settling of the inland plains are threaded through with ambivalence and contradiction, destabilized by economic and social changes, and undermined by heightened ecological and racial sensitivities. The onset of drought serves to remind us that the ecological and social dimensions of the inland have always been more closely entwined than contemporary discourses about rural Australia suggest. In these intermediate spaces of inland Australia, the physical and emotional scars of the 20th century run deep.

Notes

1 Australia's wheatlands encompass the main regions in which wheat is grown on a commercial scale for local consumption and export. Whilst the extent of the wheatlands has increased markedly during the 20th century, the main wheat-growing regions are generally restricted to the inland slopes and plains. The term 'wheat belt' was widely used in the early 20th century to describe the introduction of wheat crops to pastoral land previously used for grazing stock. By the end of the century, commercial wheat-growing was generally associated with more diversified land uses including mixed cereals and stock-grazing.

2 The origins and influence of the pioneer legend has been widely discussed. The Journal, *Historical Studies*, for example, dedicated its October 1978 issue to papers on the 'Australian Legend'.

3 Heritage conservation emerged as a distinctive nationalistic movement in Australia during the 1970s and 1980s. In 1972, the incoming Federal Government under the Prime Ministership of Gough Whitlam won office partly because of its appeal to the past as a unifying force in Australian society.

4 The statistical definition of a farming enterprise was changed in 1986-87, when the Australian government increased the threshold annual value for farms from $A5,000 to $A20,000, effectively eliminating an additional 40,000 small farms where the farmers relied upon off-farm income.

5 For example, the opening ceremony of the Sydney Olympic Games held in 2000, the Shearers' Hall of Fame established in New South Wales in 2001, and the Year of the Outback celebrated in 2002.

6 The area of land sown to crops calculated in the *ABS Year Book 2002* shows that in 1949-50 the total was 8,546,000 ha. By 1969-70 that figure had increased to 15,728,000 ha, and by 1999-2000 the total was 23,769,000 ha of which nearly half (12,168,000 ha) was sown to wheat for grain production.

7 'Scrub' (also known as 'woody weeds) is commonly used as a derogatory term to describe dense vegetation regrowth that occurs when woodland environments are cleared of trees for agriculture and grazing.

8 Immortalized in A.B. 'Banjo' Paterson's poem 'Clancy of the Overflow'.

9 The Lachlan River valley was the traditional country of the Calare or Kalar, a distinctive group of Wiradjuri-speaking people. Wiradjuri was one of the largest Indigenous 'nations' in Australia, involving more than 30 separate hordes or clans occupying much of the southern and central plains of New South Wales (Kabaila, 1996).

10 The Mabo decision in 1992 and the Wik decision in 1996 in which the High Court of Australia gave legal status to Indigenous ownership of traditional lands, acknowledging the pre-colonial presence of Aboriginal people and the continuity of Indigenous connections with particular tracts of land.

11 A portion of the Condobolin General Cemetery was set aside for the reburial of remains found in these graves (Lachlan Shire Council National Estate Programme 1981/82, 'Study of Aboriginal Places').

References

Australian Bureau of Statistics (1992), *Characteristics of Australian Farms*, Catalogue No. 7102.0, Canberra.

Australian Bureau of Statistics (1993), *The Labour Force in Australia*, Catalogue No. 6203.0, Canberra.

Australian Bureau of Statistics (2002), *Year Book*, Canberra.

Bath, T.H. (1949), 'In Tune with Nature', *The Land Farm and Station Annual*, 26 October 1949, pp. 37-8.

Bean, C.E.W. (1945), *On the Wool Track*, Angus and Robertson, Sydney.

Black, L. (1941), *Burial Trees. Being the First of a Series on the Aboriginal Customs of the Darling Valley and Central NSW*, Robertson and Mullens Ltd, Melbourne.

Bonyhady, T. (2000), *The Colonial Earth*, Miegunyah Press, Carlton, Victoria.

Carter, D. (1994), 'Future Pasts', in D. Headon, J. Hooton and D. Horne (eds), *The Abundant Culture: Meaning and Significance in Everyday Australia*, Allen and Unwin Australia Pty Ltd, St Leonards, NSW, pp. 3-15.

Commonwealth of Australia (1995), *Native Vegetation Clearance, Habitat Loss and Biodiversity Decline: An Overview of Recent Native Vegetation Clearance in Australia and its Implications for Biodiversity*, Biodiversity Series, Paper No. 6, Biodiversity Unit, Department of the Environment, Sport and Territories, Canberra.

Commonwealth of Australia (1996), *Australian Rural Policy Papers 1990-95*, Australian Government Publishing Service, Canberra.

Commonwealth of Australia (2001), *Australian Natural Resources Atlas*, vol. 2.0, http://audit.ea.gov.au/ANRA/atlas_home.cfm.

Cronon, W. (1992), 'A Place for Stories: Nature, History and Narrative', *The Journal of American History*, vol. 78, pp. 1347-76.

Curthoys, A. (2003), 'Constructing National Histories', in B. Attwood and S.G. Foster (eds), *Frontier Conflict: The Australian Experience*, National Museum of Australia, Canberra, pp. 185-200.

Fitzgerald, R. (1990), 'Writing Contemporary History in Australia', in B. Hocking (ed.), *Australia Towards 2000*, The Macmillan Press Ltd, Hampshire, pp. 65-76.

Gibson, R. (1992), *South of the West: Postcolonialism and the Narrative Construction of Australia*, Indiana University Press, Bloomington and Indianapolis.

Griffiths, T. (1996), *Hunters and Collectors: The Antiquarian Imagination in Australia*, Cambridge University Press, Cambridge.

Griffiths, T. (2002), 'How Many Trees Make a Forest? Cultural Debates About Vegetation Change in Australia', *Australian Journal of Botany*, vol. 50, pp. 375-89.

Hirst, J. (1978), 'The Pioneer Legend', *Historical Studies*, vol. 18, pp. 316-37.

Hirst, J. (1982), 'The Pioneer Legend', in J. Carroll (ed.), *Intruders In the Bush: The Australian Quest for Identity*, Oxford University Press, Melbourne, pp. 14-37.

Horton, D. (2000), *The Pure State of Nature: Sacred Cows, Destructive Myths and the Environment*. Allen and Unwin, St Leonards.

Ireland, T. (1996), 'Excavating National Identity', in *Sites: Nailing The Debate: Archaeology and Interpretation in Museums*, Paper presented to Historic Houses Trust of New South Wales Seminar held on 7-9 April 1995, Historic Houses Trust of New South Wales, Glebe, Sydney, pp. 87-106.

Kabaila, P. (1996), *Wiradjuri Places: The Lachlan River Basin*, Black Mountain Projects Pty Ltd, Canberra.

Kapferer, J.L. (1990), 'Rural Myths and Urban Ideologies', *The Australian and New Zealand Journal of Sociology*, vol. 26, pp. 87-106.

Lambert, J. and Elix, J. (2000), *Social Values of the Native Vegetation of New South Wales: A Background Paper of the Native Vegetation Advisory Council of New South Wales*, Background Paper No. 3, Native Vegetation Advisory Council of New South Wales, Sydney.

Leopold, A. (1970), *A Sand County Almanac*, Ballantine Books, New York (First Edition 1949).

Lockie, S. (2000), 'Crisis and Conflict: Shifting Discourses of Rural and Regional Australia', in B. Pritchard and P. McManus (eds), *Land of Discontent: The Dynamics of Change in Rural and Regional Australia*, University of New South Wales Press Ltd, Sydney, pp. 23-5.

Lowenthal, D. (1985), *The Past is a Foreign Country*, Cambridge University Press, Cambridge.

McKenzie, F. (1994), *Regional Population Decline in Australia: Impacts and Policy Implications*, Australian Government Publishing Service, Canberra.

Mercer, D. (1995), *'A Question of Balance': Natural Resources Conflict Issues in Australia*, The Federation Press, Annandale.

Merrett, D.T. (1977), 'Australian Capital Cities in the Twentieth Century', *Monash Papers in Economic History*, no. 4, pp. 3-26.

Mitchell, T.L. (1848), *Journal of an Expedition into the Interior of Tropical Australia, In Search of a Route From Sydney to the Gulf of Carpentaria*, Longman, Brown, Green and Longmans, London.

Nadolny, C., McMahon, S., and Sheahan, M. (1995), in P. Price (ed.), *Socio-economic Aspects of Maintaining Native Vegetation on Agricultural Land. Proceedings of a National Workshop*, Melbourne, 19 June 1995 (Land and Water Resources Research and Development Corporation, Canberra), pp. 33-4.

New South Wales Government (c.1999), *Towards a Native Vegetation Conservation Strategy for New South Wales* (brochure).

Otto, P. (1993), 'Forgetting Colonialism', *Meanjin*, vol. 52, pp. 545-8.

Powell, J.M. (1988), *An Historical Geography of Modern Australia: The Restive Fringe*, Cambridge University Press, Cambridge.

Productivity Commission (1999), *Impact of Competition Policy Reforms on Rural and Regional Australia*, AusInfo, Report No. 8, Canberra.

Ramson, W.S. (ed.) (1988), *The Australian National Dictionary: A Dictionary of Australianisms on Historical Principles*, Oxford University Press, Melbourne.

Read, P. (1988), *A Hundred Years War: The Wiradjuri People and the State*, Australian National University Press, Canberra.

Reeves, G.W. (1990), 'The Future of Australian Agriculture', in B. Hocking (ed.), *Australia Towards 2000*, Macmillan Press, Hampshire, pp. 242-55.

Robin, L. (1998), *Defending the Little Desert: The Rise of Ecological Consciousness in Australia*, Melbourne University Press, Carlton, Victoria.

Rolls, E.C. (1981), *A Million Wild Acres: 200 Years of Man and an Australian Forest*, Nelson, Melbourne.

Rolls, E.C. (1999), 'Land of Grass: The Loss of Australia's Grasslands', *Australian Geographical Studies*, vol. 37, November, pp. 197-213.

Rose, D.B. and Clarke, A. (1997), *Tracking Knowledge in North Australian Landscapes: Studies in Indigenous and Settler Ecological Knowledge Systems*, Australian National University North Australia Research Unit, Northern Territory.

Seddon, G. (1976), 'The Evolution of Perceptual Attitudes', in G. Seddon and M. Davis (eds), *Man and Landscape in Australia: Towards an Ecological Vision*, Papers from a Symposium held at the Australian Academy of Science, Canberra, 30 May - 2 June 1974, Australian Government Publishing Service, Canberra, pp. 9-17.

Seddon, G. (1997), *Landprints: Reflections on Place and Landscape*, Cambridge University Press, Cambridge, New York, Melbourne.

Seddon, G. (1998), 'Landscape', in G. Davison, J. Hirst and S. Macintyre (eds), *The Oxford Companion to Australian History*, Oxford University Press, Melbourne, pp. 374-6.

Shapcott, T. (1991), 'Poetry and place', in *Voices*, vol. 1, p. 23.

Sivertsen, D. (1994), 'Native Vegetation Crisis in the Wheat Belt of NSW', *Search*, vol. 25, pp. 5-8.

Stanner, W.E.H. (1969), *After the Dreaming*, Boyer Lectures, Australian Broadcasting Authority, Sydney.

Vanclay, F. and Lawrence, G. (1995), *The Environmental Imperative: Eco-Social Concerns for Australian Agriculture*, Central Queensland University Press, Rockhampton, Queensland.

Walter, J. (1990), 'Nation and Narrative: The Problem of General History', in B. Hocking (ed.), *Australia Towards 2000*, Macmillan Press Hampshire, pp. 77-92.

Chapter 4

Place and Presbyterian Discourse in Colonial Australia

Lindsay Proudfoot

Introduction

Despite the enduring image of 'the Bible following the Flag', the relationship between Christian missionary endeavour and 18th- and 19th-century imperial expansion was complex and far from synchronous. Missionary activity might predate formal colonization by several decades, and when the flag in question finally arrived it was by no means necessarily the national flag of the pioneer missionaries concerned. Moreover, in British settler colonies such as Australia, Canada or New Zealand, much of the aid emanating from the home Churches was directed towards the support of the colonists rather than converting the indigenous population (Porter, 1993; Etherington, 1999; Porter, 1999). Eurocentric in character, frequently supportive of and benefiting from colonial authority, Catholic and Protestant Churches might appear to be prime candidates for the role of junior partner in the assertion of hegemonic Western cultural values as part of the process of nineteenth-century European imperialism (Rotberg, 1965; Ade Ajayi, 1965; Ayandele, 1966; MacKenzie, 1999). The straightforward cultural binarism implicit in this sort of reading has given way, however, to a recognition of the ambiguous social and cultural spaces created by the European Churches during their colonial encounters overseas, reflecting the recent concern with cultural hybridity displayed by postcolonial critiques of empire generally (Comaroff, 1991; Porter, 1997; Castle, 2001). In these less clearly framed cultural spaces, the interplay between European theologies and pre-Christian cosmologies has been recognised as part of the emergence of distinctively indigenous Christian discourses (Guy, 1983; Gray 1990).

Notwithstanding their emphasis on hybridity and interchange, these *tropes* remain located within a postcolonial understanding which privileges the 'Otherness' of the colonised populations. In this chapter our concern is also with the hybrid spaces created by Christian endeavour within the cultural practice of Empire, but in this instance we focus on the disputed meanings of place and fractured collective identities which developed within a particular European settler 'faith community' as a result of their shared but contested religious practice in one particular part of Empire – South-East Australia.

The people in question are the predominantly Scottish Presbyterians who settled in various parts of New South Wales and Victoria during the nineteenth century. Australian historiography has emphasised the importance of denominational allegiance in the social and political development of the Australian colonies both before and after Federation in 1901 (Prentis, 1987; Thompson, 1998; Macintyre, 1999). Pre-eminent in this understanding have been representations of the sizeable and predominantly Irish Catholic minority as a non-anglophile 'Other', distanced – initially at least from the structures and practices of hegemonic English colonial authority by their religious affiliation, pre-modern cultural orientation, relative poverty, and perceived disloyalty. In 1868, for example, the attempt by one Irishman to assassinate the Duke of Edinburgh, in Sydney, provided apparently ready evidence for instantly suspicious sections of the Australian public of the long-standing duplicity of the Irish community as a whole (McKinlay, 1970; Keith, 1987; O'Farrell, 1993; Travers, 2000). O'Farrell argues that it was precisely because of their feelings of marginalization – however induced that many Irish migrants increasingly resorted to the agency of the Catholic Church as the major channel through which their cultural identity and interests might be mediated in the face of official suspicion. In this view, the resultant mixture of anglophobic identity politics and anti-imperial egalitarianism formed one of the main sources for the emergence of a specifically Australian sense of national identity in the later 19th century (Pawsey, 1986; O'Farrell, 1990, 1993).

By the 1870s, the dominance of Irish laity and clergy within the Catholic Church in Australia was complete. By 1891, approximately 80 per cent of the country's 228,000 Irish-born population were Catholic, and the vast majority of its Catholics were Irish (Kiernan, 2001). Much the same might also be said of the Scots and Presbyterianism. Despite the presence of a significant number of Catholics among the Gaelic-speaking Highlanders who emigrated from the Western Highlands and Islands from the late 1830s onwards, over 85 per cent of all Scots emigrating to Australia in the nineteenth century belonged to one or other of the Presbyterian Churches. Accordingly, Presbyterianism has been represented as the primary 'vehicle for the transmission of Scottish identities' in Australia as in other settler colonies in the Empire (Prentis, 1987; Harper, 2003). But not all Presbyterians in Australia were Scottish. Irish Presbyterians, predominantly from north and east Ulster, were prominent among the non-Catholic minority in the Irish migrant stream, and with their Church of Ireland neighbours, constituted perhaps 45 per cent of the Irish-born population in New South Wales in the 1840s (Proudfoot, 2003). With the increasing shift towards Catholicism displayed by subsequent generations of Irish migrants, both these Irish Protestant traditions become progressively harder to detect at a local level within the colonial record. The reasons for this are not altogether clear. The assimilation of Ulster Presbyterians into the Scottish-dominated Presbyterian Church may have been facilitated by their shared cultural background, and, more immediately, by the close links maintained between the Synod of Ulster and the Presbyterian Church of New South Wales (Prentis, 1989). Church of Ireland migrants were more polarized socially, and included significant numbers of gentry as well as tenant farmers and labourers. Many of the former appear to have been recruited fairly effortlessly into the ranks of colonial Australia's anglophile elite (Forth, 1991, 1992; McClelland, 2003).

Recognition of the social, religious and ethnic variation within the British and Irish migration streams to Australia offers a welcome advance over previous essentialist representations of some of these movements. Discussion of Irish emigration in particular has been predicated on assumptions of an ethnic authenticity based upon narrowly-defined racial origins and denominational allegiances (Proudfoot, 2003). But even this newly imagined diversity fails to capture the ephemeral localism of settler identities the ways in which their discursive social and cultural practices were mediated through place, temporarily grounding their sense of selfhood in the 'here and now' of complex and fundamentally unstable semiotic landscapes. Thus framed, the individual experience of Empire was both inchoate and conditional: inchoate because it was always developing, always in a state of 'becoming'; conditional, because the trajectory of individual experience necessarily invoked the 'Other' of hegemonic and subaltern difference against which to measure the Self.

This chapter explores these issues via an interrogation of the colonial experiences of the Reverend William Hamilton (1807-1879), one of a number of pioneer Scottish clergymen who were instrumental in establishing Presbyterianism in Australia. Hamilton arrived in Sydney in November 1837, and initially was appointed as minister to Goulburn, in southeast New South Wales. Ten years later, following the Disruption of the Presbyterian Church world wide, he felt obliged to leave, and eventually moved with his family to Kilnoorat, some 500 miles to the southwest in the Western District of Victoria. As the first Presbyterian minister to be called to this region, Hamilton was responsible for establishing a network of congregations in and around Kilnoorat, and it was to one of these, Mortlake, that he moved in 1857 and from which he eventually retired in 1873.[1] Hamilton's peripatetic career was by no means unusual among pioneer clergymen in Australia. Churches of all denominations faced the same problem of ministering to widely scattered congregations during the early years of the colonies (Griffin, 1993). His particular interest lies in the diaries and journals he kept both during his journey to Australia and subsequently at Goulburn, and the correspondence he maintained both there and at Kilnoorat with other leading Presbyterian figures (Figure 4.1).[2] These archives allow us to recover something of the local construction of Presbyterian identities in colonial Australia, and in particular, to explore the ways in which the contested Presbyterian religious imaginary – its disputed theology and practice – was mediated over time and space by the individual sense of self and place.

The discussion begins with a consideration of Hamilton's personal colonial imaginary and sense of selfhood, as these appear both before he left Scotland and in his shipboard journal and diary. This provides the context for an examination of what may be construed as the failure of his colonial imagination during his first ministerial charge at Goulburn. Here, despite, or perhaps because of his close adherence to the doctrines and practices of the Established Church of Scotland, he succeeded in alienating a significant proportion of the local Presbyterian community. The final section develops this theme of conflict and alterity by

considering the circumstances surrounding the Disruption of the Australian
Presbyterian Church in 1846. This was by no means the first schism in Australian
Presbyterianism. The discussion explores the role of Hamilton and others in these,
concluding that together, they point to the ephemeral, conditional and hybrid
nature of the 'imagined' spaces created by Presbyterian discourse in colonial
Australia.

Figure 4.1 Places Mentioned in Chapter Four

Imagining the Self, Imagining Australia

The Reverend Hamilton's departure from the Firth of Forth for Sydney, on board the *North Briton* in May 1837, was the culmination of nearly two years' indecision. Since at least September 1835, the vagaries of family circumstance and prospective clerical employment in Scotland had first encouraged and then dissuaded him in both his original plan – to become a missionary in the tropics – and in his subsequent intention to go as a minister to 'the forests of America'.[3] Throughout, Hamilton displayed a consistent, albeit retrospective, faith in Providence; a well-developed sense of self-worth, a thoroughly conventional understanding of the social order, and an occasionally surprising capacity to imagine the colonial 'Other'. Thus, he construed his failure to become a missionary as Providential preparation for his 'higher degree of authority or trust' in Canada, a task for which he thought himself uniquely well-suited. When rationalising his decision to go to Canada, Hamilton displayed a distinctly metropolitan perspective on the value of colonial human capital, as well as a profound sense of his own social status:

> ...I am more adapted than most other preachers for the ministerial duties required to be exercised in that country. There our missionaries lie among a scattered and low cultivated population. For such duties I believe I am peculiarly adapted, 1st by the habit of living with little polished society, to which I have been inured, I might almost say by my relish for solitude and the solitary contemplation of nature, 2nd ...[by] the skill in simplifying and plainly and forcibly stating doctrine which...I have acquired, 3rd by my humbleness of mind, my contentment with moderate personal accommodation, and my sympathy with men of the lowest social degree in their spiritual concerns.[4]

Three months later, in December 1835, Hamilton greeted his rejection by the Glasgow Colonial Society as candidate for a church in Nova Scotia as another sign of Divine favour, not least because in the interim he had been offered a post as assistant missionary in Greenock. Hamilton found the Society's reasons entirely convincing. The Society's secretary, Dr Burns, assured him that 'the station in Nova Scotia was not one to which I was adapted, and that it would be throwing me away to send me to it, that I was best qualified to be useful to a settled and polished congregation in a town...'.[5]

During the following year, Hamilton became increasingly confident that his future lay with a Scottish church where, in contrast to his Canadian imaginary, he felt called 'to cultivate a capacity to be useful to people of the upper classes'.[6] By February 1837, however, all this had changed, and encouraged by the Reverend Dr John Dunmore Lang, Hamilton decided to emigrate to Australia. Characteristically, however, he found reasons enough to justify his *volte face*. The lack of ministerial opportunity in Scotland, and consequent strife over 'mere trifling situations'; the great need for ministers and the 'good scope for their labours' in Australia; the 'loosening of his own attachment' to Scotland; the 'prospect of being engaged in laying the foundation of the Church of Christ in what is destined to be one of the great countries of the earth'; and, inevitably, his own sense that he had 'an adaption for missionary work such as few seem to have attained', all figured in

Hamilton's *post facto* rationalization. His new found confidence in his physical health, and his association in the enterprise with the Reverend Tait, with whom he felt 'one in mind and heart in respect of [God's] service', were an added bonus.[7]

Hamilton arrived in Sydney in September 1837, after a four month journey during which his introspective self-belief led him to construct a social imaginary of shipboard life which increasingly distanced his fellow passengers. Initially, Hamilton viewed them in terms of the moral worth of their behaviour, as measured by his own austere moral rectitude.[8] One group 'depended for their happiness on eating and drinking, card playing and jesting'; another were more serious but 'wanted habits of reading and patient application to study', while a third were the 'most sedate and serious' of all, and could be found in conversation or private study for several hours each day. But even the most frivolous, Hamilton conceded, were 'not destitute of good manners or wanting in civility'.[9] With time, however, the social distance between these groups widened, as each negotiated their own identity within the confined spaces on board ship. By June, the 'disgusting ribaldry' of 'the passengers of the lowest class' was manifestly apparent.[10] A month later and the social divisions were complete. 'The passengers have been almost entirely separated', Hamilton wrote, 'those of the coarsest class monopolizing the lower cabin...we who occupy the public cabin of the poop have no reason to regret that we are left to ourselves; we enjoy in consequence a more rational and agreeable conversation'.[11]

Agreeable though this conversation may have been, it did not inhibit Hamilton from passing judgement on his clerical companions in the poop cabin, some of whom had also been recruited by Lang. His future father-in-law, the Reverend James Clow, who was emigrating to Australia after many years' service as a Presbyterian minister in Bombay, and to whose 13-year-old daughter Hamilton became betrothed while on board, was deemed to be a 'truly Christian man [whose] talents are fair but not brilliant'.[12] His shipboard sermons, moreover, were sometimes 'little adapted to interest the feelings or command the attention of illiterate men'.[13] The Reverend Tait, whose fellowship had been so important to Hamilton before departure, and who was to become Moderator of the Synod of the Free Presbyterian Church in Victoria, was dismissed by August as jealous and worldly-minded.[14] Hamilton's severest strictures, however, were reserved for a third clergyman, the Reverend Tilley. His sermons were 'too much in the philosophical style...and ill-adapted for edification', while the man himself was criticised for consorting 'in a very improper degree with the lowest class of our passengers. It is evident that they consider him one of them [and] that they have ceased to respect him as a minister and take encouragement in their extravagances from him'.[15]

Respect of this kind was very important to Hamilton, and its perceived absence in his first ministerial charge at Goulburn was to be a source of constant dissatisfaction to him. His concern with his own and others' social status constituted an essential part of his cultural memory – those feelings of authenticity and positionality which were grounded in his Scottish origins yet continued to inflect his sense of selfhood as he travelled south and, increasingly, imagined his colonial future. Various entries in Hamilton's diary record both the 'lively anticipation' with which he and his fellow passengers discussed the Australian colonies, as well as his own sense of memory and displacement.[16] But the clearest insight into his colonial imagination is provided by the 15 resolutions he laid down for his behaviour in Australia. These covered every aspect of his imagined future life, from his dealings with his servants (who were to be closely watched, to guard them from temptation), and his relationship with Episcopalians and Romanists (to be non-confrontational, despite the numerous opportunities for debate likely in the 'present state of the colonies'), to his social demeanour and pastoral ministry. Hamilton's concern with status and respect is apparent throughout, and he clearly envisaged the continuance of his familiar social order in the colonies. He should make 'as little parade as possible in the eyes of man' and avoid 'vainglorying', relying instead on his 'natural talents and abilities' to bring him to public notice; he should 'guard against disappointment' and not seek high office, but rely instead on Providence. In his dealings with his fellow Presbyterians, he should be circumspect, and not upbraid them when they differed from him. He should give 'all men of every class a fair opportunity of approving [sic] themselves before him', treating their claims to Christianity as honest unless evidence proved otherwise; and finally, he felt it 'his duty to condescend to men of low estate and speak to them frankly and readily and take an affectionate interest in their affairs'.[17]

This last resolution echoed Hamilton's Canadian imaginary, but sat a little uneasily with the social distance he had put between himself and the 'coarsest classes' on board the *North Briton*. Moreover, it seems that by this stage, Hamilton's Australian imagination extended still further. Informed, presumably, by his reading on the colonies during the voyage, he was aware of both the growing social tensions between emancipated convicts and the free born, and of his need to inform himself of 'the habits of feeling and acting which prevail in the colony'.[18] Hamilton saw it as his duty to 'undermine and do away [with] the prejudices by which the emigrant population is separated to so great a distance from the well-doing emancipated population', while at the same time 'avoid shocking the prejudices and avoid offending the feelings of the colonists'.[19] There is little evidence, however, to suggest that Hamilton succeeded in either objective, or indeed that, construed in this way, they figured largely in his concerns once he arrived in Australia. They betokened a social awareness and breadth of vision which transcended purely religious sentiment, and Hamilton was, above all, a religious man. Whatever the apparent breadth of Hamilton's social vision of Australia *en route* to that country, when he arrived, he quickly became thoroughly immersed in a narrowly Presbyterian – and contested colonial discourse.

Memory and Practice: Placing Presbyterianism in Goulburn

The *North Briton* anchored first at Hobart, and during the vessel's brief stay Hamilton quickly established that, contrary to Dr Lang's advice, there was little likelihood of an early call to the ministry in Van Dieman's Land.[20] Encouraged nevertheless by his charitable reception, he proceeded to Sydney, where despite being assured by the Governor that the colony's need for Presbyterian ministers had also been greatly exaggerated by Lang, he was accepted by the Presbytery of New South Wales and appointed to the Argyle District.[21] Hamilton arrived in Goulburn in early October, and preached to satisfactory congregations both there and at Braidwood (50 miles to the south east) over the following weeks. At the same time he made himself useful to local Presbyterians and established acquaintance with the leading figures among them. A month later he received a nearly unanimous call to be their minister.[22]

Despite the presence locally of a significant number of Presbyterians, Hamilton was the first Presbyterian minister to arrive in Goulburn. The previous absence of Protestant ministers had, in his view, created a local indifference to religion 'bordering on heathenism', which would be all the more difficult to rectify because of the thinly scattered and partly convict nature of the population.[23] In these unpromising circumstances, he was expected to erect the formal structures of Presbyterianism as practiced by the Established Church of Scotland. In effect, he was to be instrumental, with his co-religionists, in re-materializing their Presbyterian cultural memory as part of the developing discourse of Presbyterianism in Australia. Implicit in this were issues of authority, resistance and contested meaning, and these became increasingly apparent at Goulburn as Hamilton sought to impose his own relatively austere vision of Presbyterianism on his previously unchurched congregation. The formal practices of the Church were quickly put in place. Regular services were started at Goulburn every second Sunday, and were supplemented by others held at a circuit of outstations up to 100 miles distant.[24] A Sabbath school was established, elders were appointed 'according to the Scottish model', and in March 1838, having raised £324 by subscription, the congregation resolved to apply for a government grant to erect a church with seating for 200. In further testimony to the integrity of their transmitted religious imaginary, the congregation also resolved that the new church should be run according to the 'directions and practise of the Established Church of Scotland', and that its ministers should be in communion with that Church through their membership of the Presbytery of New South Wales. In a note which probably says something about the presence of Irish Presbyterians at Goulburn, the resolution also welcomed members of the Synod of Ulster, and all other Protestants whose faith was in accord with the teachings of the Established Church of Scotland.[25]

The new church was finally completed in 1841 at a cost of £685, and by 1845 the congregation had grown to around 120 communicants. At its opening, Hamilton noted that the attendance was small, and that he made no attempt to 'disparage other Churches and laud the Church of Scotland', feeling grateful, rather, that he had been instrumental in forwarding a good work which would long survive his own ministry.[26] Like other churches erected by different denominations throughout the colonies, the Presbyterian church at Goulburn represented the material embodiment of the collective religious cultural memory of the faith community involved. How these material spaces of religious belief were *read*, both by their congregations and others, is another matter. For some of the congregation at Goulburn, their church became a place of admonishment, correction and exclusion, as Hamilton increasingly tried to impose what he considered to be the necessary and gentle discipline required to bring them into line with the Church's teaching. In his address to the Synod of Australia in 1843, Hamilton made explicit what he thought the pastor's role should be:

> It is not enough that…[ministers] simply teach truth and inculcate duty. It is also incumbent on them to confute error, and reprove sin, and to trace out and expose the manifold and ever varying devices of the devil… It is incumbent on them to remonstrate with and rebuke professed believers, who are giving way to temptations; and if any should disregard such dealing and obstinately cleave to false doctrine, or persist in sinful practise, to require them, in the most solemn manner, to repent, and penitence not thereupon ensuing, to administer those censures, which Christ has authorised as the last means to be resorted to…[27]

In the first eight years of Hamilton's ministry, 12 members of the church were excommunicating for offences ranging from fornication and drunkenness, to grog selling, 'inappropriate behaviour' and gross ignorance of the scriptures. Others, including Episcopalians and Romanists who had been attracted to the church, were denied baptism for their children on similar grounds, while some were publicly admonished for 'living in sin' prior to marriage.[28] Interestingly, Hamilton noted that the majority of his ex-communicants were 'above the lowest rank and degree of wealth; in fact the greater part of them the most able to support me pecuniarily [sic]'. More ominously for Hamilton, other members still in communion exhibited a 'false sympathy' with these miscreants, and made 'common cause with the unquestionable guilty…against whom I have set my face to reprove them'.[29]

In some instances at least, Hamilton's exercise of discipline led members of his congregation to join the Episcopalians.[30] But for others, membership of his church was clearly a more positive experience. Hamilton's own diary periodically notes a sense of individual repentance or collective faithfulness which contrasted starkly with his dealings with the ex-communicants. In 1838, for example, he baptised the children of various communicants, including those of George Gatgood, a Northern Irish Presbyterian, who although a ticket of leave man, was possessed of excellent certificates and 'seemed really well disposed'.[31] A year later, in February 1839, he felt confident that 'my people…grow in attachment, and I am gaining accessions to the number of my congregation from the Episcopalians. I have been led to hope that in one instance at least I have been spiritually useful'.[32] Ostensibly, the most positive sign of support came six years later, in October 1845, in the increasingly contested and unstable circumstances surrounding the Australian Presbyterian Church's response to the Disruption of the Presbyterian Church in Scotland. In an ultimately unsuccessful attempt to shore up his own position at Goulburn, Hamilton solicited the support of his congregation for his continued ministry. One hundred and eight people, most of apparently Scottish origin, signed a call affirming their warm approbation of the course Hamilton had pursued, both in Presbyterian politics in New South Wales in general, and as their own pastor. They expressed the earnest desire that his ministry might continue to be enjoyed by them and their children, under the direction of Divine Providence.[33]

In the event, Hamilton concluded that those who signed this second call were too dispersed to provide adequate support for him.[34] But in any case, it is clear that by this time, Hamilton's own sense of place and belonging at Goulburn was being increasingly subverted by events both there and in Australian Presbyterianism in general. It is also clear that what support he continued to receive was by now generally confined to the poorer part of his congregation; the wealthier and most influential were notably absent.[35] In short, his own reading of the abstract spaces of Presbyterian religious belief at Goulburn had been destabilized and fractured by the marginalization he felt he had experienced at the hands of some of the leading members of his congregation as well as by the Episcopalians. In an echo of the strong sense of self-worth he had displayed both in Scotland and on board the *North Briton*, this sense of personal displacement and alterity was intimately bound up with his notion of the respect due to him as a Presbyterian minister, and his dismay at the lack of material support and social recognition he received. In his first few months at Goulburn, he claimed to detect hostility in the attitude of the local Church of England pastor, and lamented the failure of leading Episcopalians to include him in their social circle, even though he considered himself to be 'more educated than they... and every bit as much a gentleman'.[36] In December 1839, not many months after he had expressed his satisfaction at the 'growing attachment of his people', he made this elision between his perceived lack of success and status clear:

> To what I am to ascribe it I cannot say, but I have been [in] every way disappointed by my reception in this country and the success of my efforts to promote the ministry and peace of the Church and the formation of an intelligent and pious congregation around. I have yet acquired little influence and won little esteem compared to what I expected and I have made little progress in converting and edifying sinners.[37]

Hamilton's sense of social alienation and clerical failure was compounded by his financial difficulties. In common with many newly arrived emigrants he found the cost of living very high, the value of money being 'not more than half what it is in Scotland'.[38] More seriously still, Hamilton soon found that he could rely neither on local Scottish Presbyterians to support him financially in a way he thought commensurate with his status, nor, under the provisions of the Church Temporalities Act, on the government to pay him an adequate stipend. The issue surfaced in March 1839, when the colonial government refused to pay him a stipend of more than £100, arguing that Presbyterians who lived more than 20 miles from Goulburn lay outside his jurisdiction. In Hamilton's view, not only did this deprive these Presbyterians of the services of a minister, but it also ensured that he was 'not provided with the means of living as becomes an educated clergyman'. Bemoaning the great backwardness of the people in supporting him, he contemplated giving up full-time ministry altogether, but concluded:

> ...I believe I ought rather to descend to live in a humbler style, though this should lead to still greater contempt...the would-be grandees have already treated me as one unworthy of their notice. Some Presbyterians I fear have turned their back upon me

because their countenancing me and receiving my services might have laid them under the obligation to pay my stipend and because Episcopacy is invested with a genteeler garb than Presbyterianism through State favour and a certain aristocratic combination.[39]

This sense of social rejection by those whose esteem he craved remained with Hamilton throughout his time at Goulburn. Ultimately, reinforced by his continuing failure to build his congregation, and against the background of the events leading to the Australian Disruption of 1846, it led him to distance himself from the whole Presbyterian community. In January of that year, notwithstanding the apparent warmth of support evinced by its poorer members, he concluded that his now falling congregation was 'evidence of ungodly opposition...the growing dislike of faithful preaching and increasing worldly-mindedness among the Scottish population of Goulburn. The people of the neighbourhood [are] no longer worthy of so much attention as I have bestowed upon them and [are] incapable of profiting from it...They impress me with the conviction that it is my duty to shake the dust off my feet and go with my message to others who have been destitute of religious ordinances and might repent, believe and be saved'.[40]

The biblical allusion to 'shaking the dust off his feet' was no doubt deliberate, and is a reminder of Hamilton's seemingly unshakable conviction in the role of Providence in his life. But there is no mistaking the complete transformation in his sense of self and place. Goulburn had been transformed in his mind from a desirable 'sphere of usefulness' to an alienating place of wasted effort. Not only so, but the cause of his failure lay not in himself, but in the unworthiness and inadequacy of those he had come to serve. So great was 'their' failure to respond adequately to 'his' message, that he was morally obliged – convicted by a sense of duty to the greater cause of the Gospel to go elsewhere. A more complete 'Othering' of his co-religionists is hard to imagine. Whatever the initial attractions of Goulburn, and Hamilton had enthused about the diverse fields of spiritual labour offered by the presence of convicts and the military as well as colonists, these had withered in the face of what he saw as his sustained social exclusion (Griffin, 1993, pp. 110-114). In this sense, the Reverend Hamilton's colonial imagination had failed him, and had proved incapable of sustaining his sense of purpose in the colonial conditions he actually encountered.

Disruption and Hybridity: Re-imagining the Place of Presbyterianism in Colonial Australia

If the Reverend Hamilton's decision to leave Goulburn was driven by his sense of his congregation's unworthiness and spiritual destitution, it was also made virtually inevitable by the part he had played in the events leading to the split in the Australian Presbyterian Church in 1846. This was itself a delayed but by no means inevitable consequence of the Disruption of the Established Church in Scotland, which had occurred three years earlier. This had been prompted by dissent over the role of the State and lay patronage in Church affairs, and led to the departure of 474 evangelical clergymen under the leadership of Dr Thomas Chalmers; these subsequently set up the Free Church of Scotland, free of all State interference (Brown, 1893). By the time news of the separation reached Sydney, Presbyterians in New South Wales had already experienced their own split and (temporary) reunification over similar issues, albeit with a strong *ad hominem* flavour.

The first Presbytery of New South Wales was established in December 1832, eight years after the first Presbyterian church, the 'Scots church', had been opened in Sydney by Dr Lang. Lang had arrived in Sydney in 1823, and had been appointed Presbyterian chaplain to the colony (Gilchrist and Powell, 1999, p. 311). As combative as he was visionary, Dr Lang devoted his extraordinarily wide talents not merely to the cause of Presbyterianism but also, increasingly, to Australia itself. Actively promoting planned emigration, he made nine voyages to Britain, sending out either additional Scots and Irish Presbyterian ministers or settlers for his various colonization schemes, including his most ambitious project for a colony of cotton growers at Moreton Bay.[41] Although these schemes were not always well founded (or funded), they reflected his vision of a future Australia largely independent of Great Britain.[42] To this end, as a member of the New South Wales Legislative Council and, later, Assembly, in the 1840s and 1850s, he successfully promoted separation for Victoria and Queensland (Gilchrist and Powell, 1999, pp. 152-58).

Until 1837, Dr Lang had a free hand in settling his clerical protégés in New South Wales and in authorizing the State to pay them a stipend (Gilchrist, 1951, p. 311). However in that year, while he was in Britain on yet another promotional tour, the Moderator of the Presbytery, the Reverend John McGarvie, applied to the colony's Legislative Council to pass The Presbyterian Church Temporalities Act. This became law in September 1837. Under its provisions, the local Presbytery became the official representative of the Church of Scotland, which previously, in its 'Declaration' of 1833, had claimed jurisdiction over all colonial Presbyteries and forbidden ministers from taking a seat in a Presbytery until they had received a call (Gilchrist, 1951, p. 311). At a stroke, Lang's discretionary authority in the disbursement of Presbyterian patronage in New South Wales had been eclipsed. On his return to Sydney in December 1837 with a further five ministers (in addition to Hamilton and the three others who had arrived in September on the *North Briton*), Lang found that not only was their placement no longer in his gift, but that he no longer had a say in the disposal of church property of which he was trustee. Like Hamilton and the others before them, the new ministers could only obtain a call, and therefore a seat in the Presbytery, once they had been accepted by it and assigned a district. Lang's reaction was immediate. Denouncing the Act as precipitate, and castigating Hamilton and his fellow travellers for their willingness to submit to the Presbytery's authority before his arrival, Dr Lang and the four Ulster ministers in his most recently arrived party withdrew, forming their own Synod of New South Wales on the 11 December, eight days after their arrival (Gilchrist and Powell, 1999, p. 243).

As the Reverend Hamilton remarked, 'Dr Lang had formed the project of a synod...long before his arrival...How much were we astonished when...we first learned from the newspapers that Dr Lang had arrived and set up a synod in opposition to the presbytery, and was prepared to wage an interminable war against that body. He created schism in a moment and he gave us no opportunity or hope to avert it'.[43] The grounds for Lang's 'interminable war' were his allegations of widespread corruption in the existing Presbytery, but these were dismissed on investigation by a Commission of the General Assembly of the Church of Scotland in 1839 (Gilchrist, 1951, p. 314). In the event, this initial schism was short-lived. In October 1840, after heated debate and while Lang was away on his next trip, to Britain and the United States, the two bodies united as the Synod of Australia in connection with the Established Church of Scotland.[44]

On Dr Lang's return in 1841, the newfound harmony proved equally short-lived. Initially accepting the union as a *fait accompli*, Lang very quickly resumed his long-standing personal invective against the Reverend McGarvie and others in the Synod. When the local Presbytery of Sydney passed resolutions against Lang for irregular behaviour, he responded in a more than usually abusive manner, describing the Synod at large as 'a synagogue of Satan', and fulminating against its continued acceptance of state aid. The inconsistency of his own position – he continued to accept his own government stipend during this time – appears to have escaped him. Nevertheless, he resigned from the Synod in March 1842 and, failing to acknowledge its disciplinary authority thereafter, was deposed from the Christian ministry in October of the same year. As Moderator, the Reverend Hamilton was in charge of these proceedings, and along with five other ministers brought to Australia by Dr Lang, concurred in his removal (Gilchrist, 1951, p. 320-30). Lang remained pastor of his now renegade congregation at the Scots church, but following his election to the Legislative Council in 1843, concentrated during the following years on his political career. In 1850, he formed his own Synod of New South Wales. Following his successful appeal in 1861 against his deposition from the ministry, this was eventually reunited with the main Synod in 1865, as the Presbyterian Church of New South Wales (Gilchrist and Powell, 1999, p. 243).

Thus in 1846, when the Scottish Disruption finally forced the hand of Presbyterians in Australia, the sequence of events leading to the schism must have appeared unnervingly familiar. As in other colonies of Scottish settlement, and despite the maverick opinions of Dr Lang, the issue of state interference in the Presbyterian Church was of far less relevance in Australia than in Scotland itself (Prentis, 1987). The Australian Synod was conscious of this, and initially, at the Reverend Hamilton's instigation, attempted to maintain a position of neutrality between the two warring factions in the Scottish Church. Hamilton thought it the colonial Church's duty 'above all things…and by all lawful means to keep united'.[45] His resolution proposing that, while acknowledging the sufferings of the Free Church, the Australian Synod should maintain communion with both it and the Established Church, was passed at the annual Synod in October 1844, but against vociferous opposition from the zealots of both parties, and with support that was diverse and potentially unstable.[46]

A year later, in October 1845, the Synod adjusted its position, and contrary to the spirit of Hamilton's proposal, passed resolutions that were more condemnatory of the Established Church than the Free. Hamilton thought this would inevitably lead to a schism, but events were overtaken by the arrival during the synod of news of the Scottish Churches' reaction to his resolution of the previous year. The Free

Church ridiculed neutrality as a 'milk and water' compromise, and demanded that the Australian Synod show the same self-sacrifice as its own members. The Established Church also rejected Hamilton's resolution, making it clear that ministers and elders adhering to it would no longer be in communion with it nor be considered its members. As the colonial government paid ministerial stipends precisely on the basis of this connection, the threat was a real one: neutrality would cost ministers their stipends and their homes.[47] The Established Church's response concentrated minds. A year later, at the General Synod in October 1846, 16 ministers reaffirmed their adherence to the principles of the Established Church of Scotland; four withdrew to form a Free Church. Hamilton protested against the Synod identifying itself with the Established Scottish Church in this way, surrendered his manse and stipend, and proclaimed himself an 'independent seceder'. Although sympathetic with the aims of the Free Church in Scotland, he regretted the actions of their supporters in Australia in promoting a local schism.[48]

If the Reverend Hamilton's experiences at Goulburn testified to the contested meanings which could attach to personal imaginings of colonial place, the Disruption and the schisms which preceded it bore witness to the wider fracturing that might occur in the collective Presbyterian religious imaginary. They also demonstrated the importance of cultural memory in the creation of colonial religious geographies. By 1840, the Established Church had withdrawn its claim to jurisdiction over the colonial Kirks, but retained its right to admonish and reprove those individuals who had been sent out as ministers by them and in communion with them (Gilchrist, 1951, p. 315). Here was institutional hybridity of a particularly ambiguous kind, made all the more so because it was projected half way round the globe. On the one hand, the Established Church refrained from interfering with the structures and practices of its colonial offshoots, though, as we have seen, these intentionally emulated it. On the other hand, it was as instant as time and distance allowed in exercising its residuary moral authority over individuals in the colonial Church. In a sense, the Established Church in Australia occupied a 'third space', independent of the Scottish Church in name, but subject to its continuing metropolitan influence; emulative of the Scottish Church's character, and constantly invoking their shared cultural memory to reshape its own Australian present (Prentis, 1987, pp. 134-46).

Such re-imagining of the present, the inchoate and conditional sense of self and place, was central to the continuing discourse of Presbyterianism in Australia. Failing in his initial objective of retaining his charge at Goulburn as an independent seceder, the Reverend William Hamilton's re-imagining took him and his family first, to his father-in-law's property just outside Melbourne, and eventually to Kilnoorat in Western Victoria. The party arrived in Melbourne in January 1847, after a nine week overland journey which has been represented by the modern Presbyterian historical imagination in heroic terms (Reid, 1959; Gilchrist and Powell, 1999). Characteristically, however, Hamilton's journal makes it clear that despite the 'losses, crosses and manifold troubles' which 'attended and severely tried' them, it was 'the impatience and insolent behaviour and language of my two hired servants...[which] tried and vexed me more than anything else'.[49] Seemingly, in the dislocating spaces of the journey, his social imaginary was challenged by the close proximity of two of the 'men of the lowest estate' he had previously claimed to cherish.

Hamilton first arrived in Kilnoorat on an extensive tour of the Western District during which he visited various communities 'with the view of ascertaining whether I could again be settled as a minister in any of the destitute neighbourhoods'.[50] He had already envisaged his ideal situation before leaving Goulburn: a district with 'a pretty dense population in a larger proportion Scottish than at Goulburn and a small nucleus of pious persons or one or two influential individuals who are sincerely interested in the cause of Christ'.[51] Writing to his wife during his tour, Hamilton reinterpreted this religious imaginary. In the relatively sparsely populated pastoral districts of Western Victoria, it was the presence of a Scottish squattocracy, anxious to obtain a minister and more than capable of paying for his services, which was the attraction. 'All whom I have seen appear anxious to obtain the services of a minister. Their distances from one another are very great and…travelling in winter must be difficult, yet the occupiers of 30 runs are considered able to contribute about £200 per annum and are likely to do so…I do not doubt that with a good deal of travelling and absence from home a minister would enjoy a good opportunity of usefulness'.[52]

Hamilton received his call in February 1847, in terms which promised him relief from 'all matters of a secular nature' and 'all due respect, submission and encouragement in the Lord', effectively offering a resolution of the two issues which had been central to his feeling of alienation at Goulburn.[53] As at Goulburn, the structures of the church were quickly put in place. Plans were laid for a church and manse, and Hamilton's preaching itinerary was agreed, effectively mapping out an imagined religious geography articulated on the runs of the Scottish squatters subscribing to the church the emplaced materialization of hegemonic social authority in the area.[54] This aptly mirrored the patronal influence of this group within the affairs of the new church. Whatever the ostensibly democratic nature of the structures put in place to ensure the accountability of the minister to his congregation via the mediation of its elders, these effectively created an oligarchy. At the annual election of the elders' committee, the same litany of wealthy Scottish run-owners reappeared: Black, Curdie, Anderson, McKinnon, Eddington, Davidson, Webster and Ross, among others.[55] Local authority within the church was disbursed among very few and intimately familiar hands.

Three years after receiving his unanimous call to Kilnoorat, Hamilton reflected on his time in the township. Once again, the elision between his sense of religious selfhood and social imaginary is made clear, but so too is the extent to which these had been re-formed in the changing 'here and now' of his life in Western Victoria. Moreover, whether because of the – for him – socially comfortable structures of hierarchical deference established within the church or for some other reason, his social imaginary appears to have been more comfortably grounded here than in Goulburn. Despite a continuance of the same material demands which had characterised life there, the 'building, planting and labouring for domestic necessities', and the same extensive 'itinerating, teaching and preaching', Hamilton felt that amid many troubles he had also experienced many comforts and pleasures. 'A liberal stipend has been paid to me, and at length I am in possession of a convenient and sufficient dwelling…The people to whom I minister with much imperfection have been very forbearing and kind, and all around have uniformly evinced a courtesy as great as I could desire'.[56] The contrast with his sense of alienation at Goulburn is profound. If Hamilton's colonial imagination had failed him there, at Kilnoorat, seemingly, he found a more congenial society in which he felt in place.

Conclusion

This chapter has explored the contested meanings of place and fractured identities which attached to Presbyterian discourse in the early years of European settlement in New South Wales and Victoria. Using the colonial narrative of a foundational figure in Australian Presbyterianism, the Reverend William Hamilton, it has demonstrated the inchoate and conditional sense of self and place which the European experience of Empire might involve. Highly motivated and confident of his own moral rectitude, Hamilton constructed a colonial imaginary in which his own sense of religious identity and imagined social status were closely entwined and mutually supportive, but which proved incapable of sustaining his sense of moral purpose in the colonial circumstances he encountered. Initially imagining his first charge at Goulburn in spiritually positive terms, his own sense of place there was dislocated by the contrary readings of others whose own meanings of place were inimical to his morally-charged aspirations and sense of self worth. Subsequently, at Kilnoorat, he encountered a collective social imaginary more in keeping with his own. Facilitated both by ethnicity and social deference, he set about establishing the material structures of his Church in a manner more in tune with the local *mentalité* than had perhaps been the case at Goulburn.

If Hamilton's encounters with Goulburn and Kilnoorat speak to the localism which lay at the heart of the European experience of Empire, his time as Moderator and role during the 1846 Disruption, remind us that this localism also mediated broader, collective experience. Each was imbricated with the other in a complex web of structure, agency and practice which extended unevenly over time and space. Thus Hamilton's concern to exercise a 'gentle discipline' over his congregation at Goulburn found an altogether more resonant echo when, as moderator, he urged the necessity of unbending spiritual discipline on the Presbyterian Church as a whole. Similarly, his unyielding conviction that the colonial Presbyterian Church should remain united in the face of the schism in the Established Church of Scotland, cost him his local material existence at Goulburn, and led to his 're-imagining' of his sense of self in place in Western Victoria. But as we have seen, by this time his personal sense of engagement at Goulburn had run its course, reminding us once and for all of the conditional and unstable nature of the European spaces of Empire.

Notes

1 William Hamilton Papers, 1835-1853 [sic], National Library of Australia, ANL: MS 2117, Private Journal 1840-1853 and later, entry for 12 January. 1873.

2 William Hamilton Papers, 1835-1853, National Library of Australia, ANL: MS2117: Diary, May-November, 1837 (hereafter, 'Diary'); Journal, 1835-1839 (hereafter, 'Journal, 1835-9'; Journal, 1838-1845 (hereafter, 'Journal, 1838-45'); Private Journal, 1840-1853 and later (hereafter, 'Private Journal'); Correspondence, 1839-1846 (hereafter, 'Hamilton Correspondence'). Kilnoorat & Darlington Presbyterian Church Board & Congregational Minutes 1847-1894, State Library of Victoria MS 11975, Call No. 2497/5 (hereafter, 'Kilnoorat').

3 Journal 1835-9, entries for September-October 1835, 4 January-22 December. 1836; Diary, 11 May, 1837.

4 Journal 1835-9, 12 October 1835.

5 Journal 1835-9, 8 December 1835.

6 Journal 1835-9, 21 January 1836, 22 December 1836.

7 Journal 1835-9, 26 February 1837.

8 Journal 1835-9, 16 April 1837.

9 Diary, 2 June 1837.

10 Diary, 10 June 1837.

11 Diary, 21 July 1837.

12 Journal 1835-9, entries for 24 July and 24 September 1837.

13 Diary, 9 June 1837.

14 Journal 1835-9, 17 August 1837; Griffin, 1993, *They Came to Care,* p. 7.

15 Diary, entries for 9 and 14 June 1837.

16 Diary, entries for 21 July, 9 August, 17 August and 21 August 1837.

17 Journal 1835-9, 8 August 1837.

18 Diary, entries for 3 and 17 August 1837; Journal 1835-9, 8 August 1837.

19 Diary, entries for 3 and 17 August 1837; Journal 1835-9, 8 August 1837.

20 Diary, entries for 5 and 18 September 1837.

21 Diary, 27 September 1837.

22 Journal 1838-45, summary entry, 27 February 1838.

23 Journal 1835-9, entries for 17 and 19 November 1837; Griffin, 1951, pp. 110-4.

24 Journal 1835-9, entries for 17 and 19 November 1837; Griffin, 1951, pp. 110-4.

25 Journal 1838-45, 14 March 1838.

26 Journal 1838-45, 4 December 1838, 16 June 1841, 23 February 1845, 2 February 1846; Gilchrist, Archibald and Powell, Gordon (1999), *John Dunmore Lang. Australia's Pioneer Republican*, New Melbourne Press, Melbourne, p. 128.

27 Hamilton, W. (1843), *The Bishop's Office: a sermon preached at the opening of the Synod of Australia, on the 4th October, 1843,* Kemp and Fairfax, Sydney.

28 Journal 1838-45, entries for 23 February, 24 March and 20 April 1845.

29 Journal 1838-45, entries for November and December 1845, 3 February 1846.

30 Journal 1838-45, 10 September 1843.

31 Journal 1838-45, 17 June 1838. 'Ticket of Leave' men were convicts who, because of good behaviour, were given their freedom on licence.

32 Journal 1838-45, 7 April 1839.

33 Hamilton Correspondence, Address to the Reverend William Hamilton MA, Minister of the Presbyterian Church at Goulburn, 20 October 1845.

34 Private Journal, 27 January 1846.

35 Private Journal, 27 January 1846.

36 Journal 1835-9, 7 March 1839.
37 Journal 1835-9, 12 June 1839.
38 Journal 1835-9, 17 November 1837.
39 Journal 1835-9, 7 March 1839.
40 Journal 1835-9, 7 March 1839.
41 Gilchrist and Powell estimate that Lang may have been responsible for sending up to 10,000 emigrants to Australia, pp. 74-5.
42 John Dunmore Lang Papers, National Library of Australia, ANL: MS 3267, Series 1, Letters 1823-1893, Box 1: Folder, Letters 1850-52: 158/195, Andrew Pringle, Tourbourie, to J.D. Lang, Sydney, 11 July 1850; 158/205, G. Rowland to J.D. Lang, 10 February 1851; Box 4: Folder, Immigration and Convicts 1850-1869: John Tweedale, Geelong, to J.D. Lang, Sydney, n/d but c. 1850.
43 Hamilton Correspondence, William Hamilton, Goulburn to Dr Burns, Paisley, Scotland, n/d but c. June-July 1838.
44 Hamilton Correspondence, William Hamilton, Goulburn, to Helen Muir, Edinburgh, n/d but c. November 1840.
45 Private Journal, 27 January 1846.
46 Private Journal, 17 October 1844.
47 Private Journal, 27 January 1846.
48 Private Journal, 27 January 1846.
49 Private Journal, 9 October 1846.
50 Private Journal, entry for January 1847.
51 Private journal, 11 January 1846.
52 Reverend James Clow Papers, 1790-1861, State Library of Victoria, MS 9570, Box 334/1[a], Original correspondence of Rev. James Clow and Mrs Margaret Clow: William Hamilton, Mount Shadwell, to Margaret Hamilton, 1 February 1847.
53 Reverend James Clow Papers, 1790-1861, State Library of Victoria, MS 9570, Box 334/1[a], Original correspondence of Rev. James Clow and Mrs Margaret Clow: J. A. Webster *et al.* to William Hamilton, 15 February 1847.
54 Kilnoorat, entry for 20 April 1847.
55 Kilnoorat, entries for 20 April 1848, 20 April 1849, 26 April 1850, 20 April 1852.
56 Private Journal, 30 April 1850.

References

Ade Ajayi, J.F. (1965), *Christian Missions in Nigeria, 1841-1891: The Making of a New Elite*, Longman, London.
Ayandele, E.A. (1966), *The Missionary Impact on Modern Nigeria, 1824-1914*, Longman, London.
Castle, G. (ed.) (2001), *Postcolonial Discourses. An Anthology*, Blackwell, Oxford.
Comaroff, J. (1991), 'Missionaries and Mechanical Clocks: An Essay on Religion and History in South Africa', *Journal of Religion in Africa*, vol. 71, pp. 7-17.
Etherington, N. (1999), 'Missions and Empire', in S. Macintyre (ed.), *A Concise History of Australia*, Cambridge University Press, Cambridge, pp. 303-14.
Forth, G. (1990), 'The Anglo-Irish in Australia. Old World Origins and New World', in P. Bull, C. McConville and N. McLachlan (eds), *Irish Australian Studies. Papers delivered at the Sixth Irish-Australian Conference*, La Trobe University, Melbourne, pp. 51-62.

Forth, G. (1992), '"No Petty People": The Anglo-Irish identity in colonial Australia', in P. O'Sullivan (ed.), *The Irish World Wide. History, Heritage, identity. Volume Two. The Irish in the New Communities,* Leicester University Press, Leicester, pp. 128-42.

Gilchrist, A. (ed.) (1951), *John Dunmore Lang. Chiefly Autobiographical 1799 to 1878,* 2 vols, Jedgarm Publications, Melbourne.

Gilchrist, A. and Powell, G. (1999), *John Dunmore Lang. Australia's Pioneer Republican,* New Melbourne Press, Melbourne.

Gray, R. (1990), *Black Christians and White Missionaries,* Yale University Press, New Haven.

Griffin, G.M. (1993), *They Came to Care. Pastoral Ministry in Colonial Australia,* JBCE Books, Melbourne.

Guy, J. (1983), *The Heretic; A Study of the Life of John William Colenso, 1814-1883,* University of Natal Press, Johannesburg.

Harper, M. (2003), *Adventurers & Exiles. The Great Scottish Exodus,* Profile Books, London.

Keith, A. (1987), *The Fenians in Australia,* University of New South Wales Press, Sydney.

Kiernan, C. (2001), 'The Irish Character of the Australian Catholic Church', in P. Jupp (ed.), *The Australian People. An Encyclopedia of the Nation, Its People and Their Origins,* Cambridge University Press, Cambridge, pp. 459-63.

Macintyre, S. (1999), *A Concise History of Australia,* Cambridge University Press, Cambridge.

MacKenzie, J.M. (1999), 'Empire and Metropolitan Cultures', in A. Porter (ed.), *The Oxford History of the British Empire. Volume III. The Nineteenth Century,* Oxford University Press, Oxford, pp. 270-93.

McClelland, I. (2003), 'Worlds apart: the Anglo-Irish gentry migrant experience in Australia', in O. Walsh (ed.), *Ireland Abroad. Politics and Professions in the Nineteenth Century,* Four Courts Press, Dublin, pp. 186-201.

McKinlay, B. (1970), *The First Royal Tour 1867-1868,* Rigby Limited, Sydney.

O'Farrell, P. (1990), *Vanished Kingdoms. Irish in Australia and New Zealand,* New South Wales University Press, Sydney.

O'Farrell, P. (1993), *The Irish in Australia,* New South Wales University Press, Sydney.

Pawsey, M. (1986), 'Aliens in a British and Protestant Land: Early Culture Clashes in Victoria', in C. Kiernan (ed.), *Australia and Ireland 1788-1988. Bicentenary Essays,* Gill and Macmillan, Dublin, pp. 72-87.

Porter, A. (1993), 'Religion and Empire: British Expansion in the Long Nineteenth Century, 1780-1914', *Journal of Imperial and Commonwealth History,* vol. 20, pp. 370-90.

Porter, A. (1997), '"Cultural Imperialism" and Protestant Missionary Enterprise, 1780-1914', *Journal of Imperial and Commonwealth History,* vol. 25, pp. 367-91.

Porter, A. (1999), 'Religion, Missionary Enthusiasm, and Empire', in A. Porter (ed.), *The Oxford History of the British Empire. Volume III. The Nineteenth Century,* Oxford University Press, Oxford, pp. 222-46.

Prentis, M.D. (1987), *The Scottish in Australia,* AE Press, Melbourne.

Prentis, M.D. (1989), 'Pioneer Irish Presbyterian Clergy in Australia, 1832-1858', *Church Heritage,* vol. 6, pp. 73-85.

Proudfoot, L. (2003), 'Landscape, Place and Memory: Towards a Geography of Irish Identities in Colonial Australia', in O. Walsh (ed.), *Ireland Abroad. Politics and Professions in the Nineteenth Century,* Four Courts Press, Dublin, pp. 172-85.

Rotberg, R.I. (1965), *Christian Missions and the Creation of Northern Rhodesia, 1880-1924,* Princeton University Press, Princeton.

Thompson, R.C. (1998), *Religion in Australia. A History,* Oxford University Press, Melbourne.

Travers, R. (2000), *Henry Parkes. Father of Federation,* Kangaroo Press, Sydney.

2.1 Post Mortem hostility: the gravestone near Carrickmacross of William Steuart Trench, agent on the Shirley, Bath, Landsdowne and other estates, was destroyed by a hostile local community in 1873, a year after his death. Following restoration in the 1970s, it was again dismantled under cover of darkness (Author's Collection).

2.2 Shirley's mock Tudor mansion house erected near Carrickmacross in the late 1830s. At annual dinners for the tenantry in its Great Hall in the mid 19th century, toasts to the Queen were drunk and lectures on frugality and moral improvement delivered by the landlord (National Library of Ireland).

8.1 W.F. Massey, Prime Minister of New Zealand circa 1916 (S.P. Andrews Collection, Alexander Turnbull Library, Wellington, New Zealand).

8.2 W.F. Massey addressing New Zealand machine gunners in France at Bois-de-Warnimont in 1918 as part of one of his ministerial visits to the troops (RSA Collection, Alexander Turnbull Library, Wellington, New Zealand).

8.3 'Peace Prosperity Empire'. Reproduced from *Quickmarch*, the Returned Services Association journal. The cartoon was originally drawn at the end of the Boer War but considered even more apt when it was reprinted in January of 1919 (Alexander Turnbull Library, Wellington, New Zealand).

8.4 Nurse Schaw was one of a handful of former Army Nurses to ballot for land under the Discharged Soldier Settlement Scheme securing a four acre section in 1919 on which she attempted to raise poultry (Author's Collection).

Chapter 5

Irishness, Gender and Household Space in 'An Up-country Township'

Di Hall

Introduction

St Patrick's Day 1863 was celebrated throughout the gold mining district of Stawell in the west of the colony of Victoria, with a race day at the rural village of Glenorchy, a cricket match in Stawell itself and grand ball in the neighbouring town of Ararat.[1] At the ball, under decorations consisting of 'a large harp, which was suspended from the ceiling and ornamented with shamrocks', William O'Callaghan, Irish-born publican and aspiring politician, made a long speech on Irish history and nationalism. The speech ended with toasts and the last was 'To the Ladies', then everyone began to dance.[2] Although both men and women were present, and indeed both were considered necessary for the dancing to be a success, the Irish identity expressed in the dining room of the hotel was masculine as well as nationalistic.

For William O'Callaghan and his companions, their identity as Irish men was expressed and reflected, at least in part, through the formation of the culturally specific Irish space of the St Patrick's Day Ball. These narratives of publicly constituted Irish spaces will yield little information about the historical formation of Irish women migrants' identities and their spatial expression. Although they were undoubtedly present at such events, they were rarely included in the descriptions of such Irish spaces as St Patrick's societies and parades or fund raising events for Irish nationalist causes.

Feminist scholars have investigated the gendered nature of space in many different social and historical contexts. One of the outcomes of this research is the recognition that space is socially constructed and at the same time social relations are spatially conceptualized in ways that are contingent on specific historical contexts (Massey, 1994; Morin and Berg, 1999). Following the argument of scholars such as Probyn, who call for attention to the 'material contexts which allow and delimit our individual and collective performance of selves' (Jacobs and Nash, 2003, p. 279) and (Probyn, 2003, p. 291), analysis of the spatial expression of Irish women's identity needs to be centred on the places where they lived, worked and socialized. The material context of most Irish women in colonial Australia (as elsewhere) was the household, a site that feminist geographers have long identified as a productive unit of analysis (McDowell, 1992).

In this chapter I will analyse the gendered spatial expressions of identities of the Irish women who lived and worked in the Royal Mail Hotel, Glenorchy, around 1863. This analysis of localized and particular gendered spaces reveals the complexity and fluidity of Irish women's identities as they passed through their life cycle and moved or attempted to move into different socio-economic groups. While the events I analyse are particular, the contexts of domestic service, marriage and households, were commonplace for many Irish-born women in colonial Victoria. It is in detailed investigation of local circumstances that the great diversity of experiences of Irish-born women in colonial Australia is revealed.

Irish Women in the Australian Colonies

Research on Irish women migrants in colonial Australia has increased over the past 20 years, with detailed analysis of immigration statistics, demographic patterns and admissions to prisons, mental asylums and benevolent homes (Jackson, 1984; Morgan, 1989; Campbell, 1991; McClaughlin, 1991; Haines, 1996; Rule, 1996; Fitzpatrick, 1998; McClaughlin, 1998; Rule, 1998; Gothard, 2001, 2001). These studies have successfully demonstrated that the migration stream to the Australian colonies included a very high proportion of single young Irish women, many of whom migrated within kin or friendship networks, and worked as domestic servants once they arrived. They married relatively early and about half of them married non-Irish men and dispersed throughout the colonies.

Feminist scholars have investigated 19th-century Australian colonial space in gendered terms by analysing how the bush was gendered masculine in art, literature and myth, while the interior, private, domesticated spaces of house, verandah and garden were gendered feminine (McPherson, 1994; Rowley, 1994). The land was also associated with gendered danger, a danger that was acute and threatening for women, whilst allowing men freedom and opportunity (Pickering, 2002). These analyses, while fruitful for focusing attention on the gendered descriptions of the mythic and iconic spaces of the Australian bush, depend in their formulation on the dichotomy between private/public spaces. Over the past ten years or so there has been increasing pressure to modify or indeed abandon this public/private model as it has been shown that public/masculine *versus* private/feminine works well for only a small number of women, particularly white middle-class 19th-century women (Blunt and Rose, 1994). Understanding women's experiences in colonial settings as being 'confined' in the private sphere has also been criticized by Mills for not allowing for the complexity of colonial gendered spatial relations (Mills, 1996). She discusses the increased sense of freedom that many white English women travellers felt when they were in colonial spaces, giving them the opportunity to move about physically that they did not have back in the metropolis. In this analysis, viewing white women's experiences through a lens of 'confinement' to domesticated spaces does not allow for this enjoyment of freedom. Moreover, the analysis of the domesticated spaces of house and garden does not adequately take into account the heterogeneity of the women who lived in colonial settings. While some middle-class white women saw themselves confined to the safety of the homestead fence, there were many other women in the Australian colonies for whom there was no homestead fence in which to be safely enclosed.

There has been increased scholarly attention paid to the roles of white women in the colonization process in 19th-century Australia, with geographers in particular analysing how white women facilitated and engaged with the colonization project, alienating and excluding indigenous women and men in the past and the present (Johnson, Huggins, and Jacobs, 2000). These analyses of white women in colonial spaces have yielded productive ways of thinking about constructions of 'whiteness' within empire and colony, however the category of 'white' or 'British' in the Australian context is usually considered to be a homogenous one. This is also a feature of contemporary and historical 'British' identity in Britain, an identity that obscures the heterogeneity of the white women who are subsumed in it. Bronwen Walter has argued that the silencing of Irish women in definitions of Britishness stems at least in part from their close identification with domestic service in the 19th century, where their silent and invisible labour was essential for the creation of the idle middle-class white woman (Walter, 2001). In the history of the Australian colonies, the strong correlation between Irish migrant women and domestic service means that they have also usually been silenced and rendered invisible when female spaces have been analysed. These female spaces have usually been middle-class spaces or those of pioneer women battling on their own whose 'Britishness' or 'Australianness' has not been questioned. Women's ethnicity is assumed to be 'British' or 'Australian' unless it is demonstrably other. This point is admirably demonstrated in the short story by 19th-century Australian-born author Barbara Baynton. In 'Billy Skywonkie' the reader's and the male characters' assumptions of the main character's whiteness are exposed when her Chinese ethnicity is revealed. (Baynton, 1902) While subsuming all white women into the category of 'colonizing British' correctly signals that they benefited from their whiteness at the expense of indigenous, Islander, Indian and Chinese men and women, 'white' women were not homogenous in either ethnic or class terms. The largest group of non-English white women were Irish.

The ethnic and racially ordered categories of the 19th-century census statistics did recognize differentiation within 'white' or 'British' women. When the various censuses were conducted in the 19th-century Australian colonies, the categories into which people were collated were those that were considered to be politically significant in terms of policy. In this way indigenous people were specifically counted in every district (Watts, 2003). Among immigrants, Chinese were differentiated while the Irish, Welsh, Scots and English were counted separately under the heading of British, at the same time that the Cornish were subsumed within the category of English (Payton, 2001). These census results mean that it is possible to count numbers of women born in Ireland and these statistics, combined with vital registration and immigration data, mean that demographic and family linkage studies can and have been undertaken of individuals and groups.

Nineteenth-century Irish women migrants cannot themselves be considered an homogenous group. To point out very stark differences: an illiterate Catholic 15-year-old from rural County Clare who worked as a domestic servant in a

Melbourne suburb in the 1850s before marrying a Lancashire-born shopkeeper would have constructed and experienced her identity as an Irish-born migrant in a different way from a Protestant woman from County Fermanagh, who migrated with husband and extended family to rural Queensland in the 1860s. Life experiences prior to migration, class, socio-economic opportunity as well as experiences over a lifetime once in the colonies, shaped and reshaped identity, making it multi-layered and changeable over time and place. Yet although these women were not the same and did not live the same experiences either in Ireland or in Australia, they were viewed by many employers, commentators and fellow migrants as having the same characteristics, as being 'Irish'. When Mary Honan, a working-class married woman, was allegedly raped on the gold fields in Stawell in 1863 she testified that her attackers had told her 'You are Irish' while they assaulted her.[3] Her identity as an Irish woman was significant enough for these men to use it as a term of abuse. So these women's 'Irishness' which they may have only perceived and experienced on leaving Ireland was also a part, but only a part, of their identity in the Australian colonies.

It is in the analysis of localized experiences that the similarities of Irish women's identities and their differences from each other and from other women and men become evident. Massey calls for the investigation of 'the variable construction of gender relations in different local-cultural space/places' (Massey, 1994, p. 178) that point to the social construction of identity. By examining the local and the particular, the complexities of women's spatial construction of identity becomes evident in ways that are masked by aggregated statistical analysis. Letters written by Irish emigrant women, for example, have productively revealed the diversity of their experience and expectations in the colonies. When Isabella Wyly wrote to her sister Matilda in Newry, County Down, from Adelaide after a seven year silence in 1856, she described the fate of various family members and then wrote 'What changes, and I suppose there has been just as many at home. It does not seem home to me now...' (Fitzpatrick, 1994, p. 113). When women such as Isabella were writing these letters they were consciously thinking of themselves as part of a network based on a particular place in Ireland, yet in their everyday lives this place was no longer 'home'. It is in the analysis of these colonial homes that the spatial construction and expression of identities of Irish-born women can be discerned.

'An Up-country township'

In Glenorchy, in early May 1863, the tensions that had been simmering in the Royal Mail Hotel overflowed into a violent and public altercation between Honora Jenkins, the wife of the publican, and her servant, Martha Jane O'Kane. Martha O'Kane testified that on the Sunday morning she had entered the upstairs bedroom to waken the nursery maid. In the room were also sleeping Honora, the 13-year-old nursery maid, Honora's baby son and Honora's 10-year-old step-daughter. When Martha opened the door, she alleged that Honora yelled at her, called her 'frosty face', got out of bed and threw her to the floor calling her a whore. The following night, Martha was sitting sewing a child's dress when Honora objected to her presence in the room, variously described as the tap room, the parlour and Honora's private room. Five of the neighbours (four women and a man) then rushed in and pulled Martha by the hair into the street, ripping her clothes off and kicking her repeatedly. Various insults and accusations were also thrown about in the course of the fight, including that Martha was a whore and a harlot. Martha then sued the five neighbours for assault and, in separate court actions, she sued two of the women for using insulting language and sued Honora for assault. One of the women, Bridget Fitzgerald then counter-sued Martha for perjury. The results of all these court actions were that the four women who had allegedly assaulted Martha were fined while Honora Jenkins was bound over to keep the peace and the other two actions were thrown out of court.[4]

The township of Glenorchy had begun when a bark and slab hut hotel was established on the road between Melbourne and Adelaide by John Gleeson in the 1840s. To cater for the non-alcoholic needs of travellers, his daughter Bridget and her husband Scots-born Robert Jenkins ran a general store, and Limerick-born couple Matthew and Honora Spellecy were invited to open a blacksmith forge. The little settlement serviced both the road traffic and also the recreational needs of the workers on the pastoral stations surrounding it (Kingston, 1989). By 1863, Gleeson had moved on and his daughter was dead, however Jenkins had stayed and prospered, rising from storekeeping to building and running the biggest hotel in town, the Royal Mail, described at the time of its completion as more substantial than many of the squatters' homesteads (Wallace, 1888).

Contemporary descriptions of Glenorchy emphasise the masculine gendered spaces and activities that predominated in the township. When John Wallace arrived there in 1853, there were only six married couples living in the town, a circumstance that he felt exemplified the township's rough nature (Wallace, 1888). The author Marcus Clarke lived on a nearby pastoral station during 1865-67 and used Glenorchy as the backdrop for several of his short stories published in the 1870s. His most explicit description of the township is as the fictional Bullocktown in 'An Up-country township' (Clarke). Here women are only mentioned as an essential (and largely missing) ingredient to masculine recreation.

> It was not often that we had amusement …in Bullocktown. Except at shearing time when the 'hands' knocked down their cheques (and never picked them up again) gaiety was scarce. Steady drinking at the Royal Cobb and a dance at Trowbridge's were the two excitements. The latter soon palled upon the palate, for, at the time of which I write, there were but five women in the township, three of whom were aged, or as Wallaby said, 'broken-mouthed crawlers, not worth the trouble of culling'. The other two were daughters of old Trowbridge, and could cut out a refractory bullock with the best stockman on the plains. (Clarke, 1976, p. 572)

For Clarke and his fictional depiction of Glenorchy, the unattractiveness of the few available women is signalled by their lack of femininity and their association with the animals that everyone depended on for their livelihoods.

Yet neither Wallace nor Clarke were seeing all the women who lived and worked in Glenorchy. For Wallace, the only visible women in the social spaces of Glenorchy were married women of his own class. Clarke only saw unmarried women available to entertain single stock men. Neither men saw (or commented on) the women who lived and worked in Glenorchy providing essential domestic services in hotels, shops and individual houses. The 1861 census recorded 54 females and 82 males living in the township of Glenorchy. While no age breakdown is given in the figures, the population statistics for the larger Glenorchy Electoral District suggest it is reasonable to assume that about half the females and a quarter of the males were children.[5] Certainly in 1863 there were at least 20 women living in the township of Glenorchy and of them 13 had probably been born in Ireland.[6] This concentration of Irish-born women was considerable in such a small township. In 1863 there were only 20 rate payers, with 14 houses of varying sizes, one general store and three hotels, of which Jenkins' Royal Mail was by far the largest. In such a small settlement, there were few opportunities for paid work for women. There would have been some call for domestic help with small children, as there was an average of 12-15 births a year registered at Glenorchy for the village and the rural region that it serviced. This need would mostly have been met by very young girls, such as the 13-year-old nurse who lived at the Royal Mail in 1863.[7] For a few other women, there was shop keeping, Honora Spellecy ran the general store, where she was rumoured to have made a fortune cashing cheques for a healthy commission (Wallace, 1888). In the 1861 census, occupations are given for women in the Glenorchy Electoral district, comprising the rural area around Glenorchy. Of 120 adult women, three worked in hotels, there was one needlewoman, one schoolteacher, 17 domestic servants and 18 pastoral workers. The rest of the women were counted as unemployed or wives. So there were considerably more women in Glenorchy than either Wallace or Clarke recorded, most of them were not of the social class to be noticed by the authors or were engaged in domestic service where they were rendered invisible and silent by their work as servants or their status as wives of labourers. Apart from work on the three pastoral stations based around Glenorchy, the only paid employment for domestic service was in homes of other residents or in the three hotels, which did a very lucrative trade servicing both pastoral workers and travellers on the road between Stawell and Adelaide.

In 19th-century rural Australian communities, hotels were important not only as places that sold alcohol, and accommodation for travellers and permanent residents, but also as providers of communal rooms where civic, social and legal activities were conducted. Publicans were required by law to live on the premises and there had to be accommodation, food and stabling provided for the public that was separate from the publican family's private rooms. Women were essential to the success of hotels in most places in the colonies, providing the labour that ensured the provision of accommodation and food as well as working in the bar. Many women held publicans' licences themselves, while others ran the hotel with their husbands as licencees (Wright, 2003). Successfully running a hotel brought publicans and their families into contact with the civic and social hierarchies of communities and was often used as a way for Irish men and women to establish themselves both financially and socially in local communities. In the mining town of Stawell, for example, in 1870 Anne Nihill and Johanna Matheison were two Irish born licensees and at least another six hotels in the town were run by Irish men and their wives.[8] Running these hotels was a means to social and financial security for many, with men such as George Jenning and Patrick D'Arcy, becoming significant public figures in Stawell as well as investing the money they made in hotels in land from the late 1860s.[9] In Glenorchy as well, hotels were a good investment: the Spellecy family, for example, diversified from blacksmithing into hotel keeping, building the Glenorchy Hotel in 1859 (Wallace, 1888).

Robert Jenkins was upwardly socially mobile, going from hawking goods to pastoral workers in the 1840s to owning the Royal Mail less than 20 years later. The Royal Mail Hotel was the largest building in Glenorchy with six bedrooms and three sitting rooms.[10] Robert was doing well enough at the hotel business to also establish a small but valuable portfolio of other property, including four separate allotments of land and four houses, with a fifth being built at the time of his death in 1875.[11] Since at least 1859 he had been employing female servants to help with the household labour of raising small children and running the hotel and store.

All the adults living at the Royal Mail Hotel constructed the meanings of the spaces they inhabited in ways that both reflected and formed the multi-layered and changing nature of their individual identities. Elements of these identities – their place of birth, class background and religion – were recorded in official statistics and give some indication of the ideological, social and material contexts in which they were living in 1863. Robert Jenkins was born in Glasgow where his father had been a coal master. He had migrated to Victoria in 1841, and soon after married Bridget Gleeson, daughter of John Gleeson the publican, settling in Glenorchy around 1844. Six years later Bridget had died and he had married his second wife, Emma Ryan, in Geelong. Emma had arrived from County Tipperary in 1847 and after her death in 1861, Robert married, for a third and last time, Honora Kenealy, who had been employed as a servant in his hotel.[12] He surrounded himself with family as well as friendship networks: his nephew worked for him, and his sister lived in Glenorchy from the early 1860s.[13] Honora was a Limerick-born Catholic woman in her early thirties. She had settled in Glenorchy as part of a network that included her sister, Ellen, who was also a servant living in the village and their brother who lived in Stawell. The servant who succeeded her at the Royal Mail was Martha O'Kane. Martha was a young Belfast-born woman who had migrated as an unassisted migrant in 1859, with her two sisters and an infant niece. Martha described herself on arrival as a housemaid and as able to read and write.[14] She was not living near her family when she was working in Glenorchy, however within five years another of her brothers and her mother had joined her in Stawell.[15] While Martha indicated that she could read and write on her arrival, Honora could not sign her name in 1861[16] This difference probably reflects Honora's rural background and Martha's urban opportunities. The class backgrounds of Honora and Martha were typical of Irish Catholic women in the Stawell/Glenorchy area. Of the 91 marriages recorded between 1858 and 1883 in St Patrick's parish church in Stawell (which included Glenorchy) 44 women were servants (including a few laundresses and cooks), and 24 were housekeepers, meaning that 75 per cent of these Irish-born Catholic women classified themselves as in domestic service.[17]

The gendered spatial and material contexts of women's lives changed over the course of their lives, most notably with changes of marital and employment status. In 1861 when Honora was the servant at the Royal Mail, the use of space and the construction of the identities of servant and mistress were, at least on the surface, uncontested. By this time Emma Jenkins' private and public use of alcohol and its associated illnesses were defining the final years of her life, and shaped how her identity was constructed by the women of Glenorchy. In the inquest into her death and the death two years earlier of her baby, she was described as being in bed throughout the day and often insensible with alcohol. Her physical needs were attended to by local women who came into the bedroom to help her, but she herself is not described as going out at this time, though she had been seen in the streets of the village while drunk on previous occasions. Honora Kenealy described how she had gone in and out of the bedroom to check on Emma and serve her tea, and various women of the village came and went in order to nurse Emma and her baby. Robert, Emma's husband and Honora's, does not have a place in the domestic and intimate spaces of the death of his wife or baby – these spaces were gendered female. Robert was in the storeroom, masculine working space, on both occasions and had to be called out to interpret the physical signs of death and to summon the authorities to confirm the diagnosis of the women who attended both Emma and the baby.[18]

Alcohol use and abuse by women in 19th-century Australia was much commented upon at the time, mostly as evidence for the need for control urged by the temperance movements. The amount of alcohol consumed by women is difficult to gauge and historians have tended to approach the topic with the assumption that women did not drink. This means that when alcohol consumption has been calculated, statistics for the gross production and importation of alcohol has been compared with male population figures (Dingle, 1980). Women have been presumed to be relevant to the statistical analysis of alcohol use only as agents for sobriety, even though as Wright has argued, they were involved in large numbers in the sale and consumption of alcohol (Wright, 2001, 2003). Women who drank in public were treated with suspicion by newspaper writers, police and magistrates, and were liable to be charged with drunkenness and public disorder offences. Many of these women were Irish (McClaughlin, 1996). Davis has postulated that in England Irish women were disproportionally more likely to be described as being involved in public brawls and drunkenness because of their differentiation from 'English' women by their Irishness, itself often signalled by domestic service (cited in Walter, 2001, p. 99).

During the course of the 19th century, women's use of alcohol became more unacceptable, although it was those women who drank in public who were perceived as the farthest removed from civilized society (Warsh, 1993). Many Irish migrant women working as domestic servants did not have access to private places in which to drink and this may have exacerbated the public perception of the high level of drunkenness of Irish women. Such public disapproval led many women with access to privacy to drink in unmonitored spaces like bedrooms. Emma Jenkins, though she was known to be sometimes drunk in public, was described as being drunk in the privacy of her bedroom throughout the day. Other Irish-born women whose husbands had achieved some degree of material success also retreated to these intimate domestic spaces in order to drink unnoticed. Elizabeth Levy was the Irish-born wife of Patrick D'Arcy, the successful Stawell publican, and died aged 30 from an alcohol related disease after living in seclusion for some years. Her husband testified that she had been drinking for at least four years.[19] Women who drank were often considered to have lost their reason and be at the same level as animals because they could not control themselves. The Glenorchy woman who nursed Emma Jenkins and her baby used this imagery when she saw Emma's dead infant, and yelled that she was 'a beast', blaming her and her use of alcohol for the baby's death.

Honora Kenealy's circumstances changed with the death of Emma and her marriage shortly afterwards to her employer, Robert Jenkins. Her public role was now that of wife of one of the richest men in the village and employer of servants, rather than as a servant herself. In the popular imagination, in the press and cartoons in Australia (and England and United States), domestic servants were usually depicted as incompetent Irish farm girls, while their mistresses were portrayed as English, middle class and urban (Hamilton, 1993). Yet the reality was more complex than the straight-forward binary oppositions used by the male writers of popular press. Typical Irish migrants at this time were young, single females who stated their occupation as some sort of servant and migrated under one of a number of government-funded schemes (Haines *et al.*, 1998; Gothard, 2001). It was recognized at the time that most of these women, when they married and set up homes of their own, employed at least one servant themselves to help with the labour of raising small children in harsh conditions (Gothard, 2001). So contrary to the stereotypes in the contemporary press, there were large numbers of Irish-born women employers, like Honora, who had experience working as servants. There was a history of domestic service in middling and strong farming, as well as middle-class households in 19th-century Ireland. In these households the single female servant was often a young girl who came to service from farming households of similar class background and worked temporarily in service prior to marriage (Luddy, 2000). The roles of servant and mistress were thus differentiated by age and marital status, rather than by differences in class backgrounds. Servants and mistresses had to share the domestic spaces in which they lived and worked, meaning that a delicate balance had to be achieved so that the social differences were visibly maintained. In this way the identity of the servant was often spatially constructed by her relegation to the 'working' areas of the household – kitchen, laundry and nursery – and the identity of the mistress was defined by her occupation of the leisured part of the house – the parlour or bedroom. While the boundaries between these household spaces were necessarily fluid, especially in small farming households, it was in the interests of the mistresses in particular to ensure that separate identities continued to be expressed in these spatial ways. This must have been much harder to achieve when the mistress herself was so recently a servant in the very household in which she now employed domestic help.

The complexity and ambiguity of Honora's changed identity was demonstrated clearly by the way that space was constructed and contested during her dispute with Martha. In less than two years since her marriage and adoption of the role of employer, her position was challenged over the use of both private and public household spaces. She was sleeping in the private quarters with her children and the nursemaid, rather than with or near her husband. She resisted the efforts of Martha to enter this space to wake the other servant and asserted herself as the injured party in what was obviously a long running dispute, calling Martha a "whore" and "frosty face", and physically removing her from the private room into the more public space of the hotel corridor. Not only did Martha occupy sleeping quarters downstairs, in the vicinity of Robert, but she also shared with him the important role as keeper of the household and hotel keys. As wife and employer this should have been Honora's job. As Martha controlled the food and supplies, Honora had lost authority, privately at least, over these important sites of household power that denoted her role as mistress of the hotel.

In the reports of the court cases, it is the use of leisured space by Martha that triggers the major altercation. She was sitting on the sofa in the parlour or private room and sewing a dress for one of the children of the family. While it is easy enough to interpret the cultural meaning of the parlour in private homes, hotel parlours were more ambiguous spaces. In some hotels they were definitely only for the family, while in others 'parlour' was the code word for the slightly more private bar where women drank and socialized. In private homes, parlours were the sign of respectability by the mid- to late-19th century. The arrangement and furnishing of the one good room in the house, which was not for every day use, but for display, was a feature of many houses in Australia and elsewhere (Lawrence, 1982). Though the hotel parlour was more public than a parlour in a private house, it was still the 'good' room where more riotous masculine behaviour, typical of the public bar, was not tolerated (Wright, 2003, p. 112). The parlour of the Royal Mail Hotel was such a room, furnished with different and more expensive furniture than the rest of the hotel.[20] For Martha to be in the parlour was doubly provocative, as she was both seated and sewing, engaged in middle-class feminized leisure activity in a semi-public/semi-domestic space, when she was the servant not the mistress. Her behaviour was interpreted by the women of Glenorchy as provocative and insulting. To emphasise her use of the culturally meaningful space of the mistress, Martha told Honora that the room was 'her place that night' that Honora was not her mistress, that she was a 'thing'. Honora understood this as a challenge to her authority and her identity as mistress in the hotel, and probably in the affections of her husband. She responded by physically removing Martha out of the private space of the parlour and into public view on the verandah. Here her neighbours supported and encouraged her. A witness reported that 'Mrs Temple she said to Mrs Jenkins, go into your house and defend yourself' and that 'Mrs Jenkins was arguing with Martha and telling her to go out of her private room'. Martha's actions were clearly recognised by the women of the township as challenging Honora for the right to use the parlour, and which undermined Honora's identity as a both a married woman and an employer of servants.

Each woman asked for and received help in the dispute over contested space at the Royal Mail. In their use of networks to support their positions the women used strategies that were common in different areas of Ireland, and their actions can be usefully analysed by recognising their Irishness. Martha and Honora were from very different environments in Ireland – urban Belfast and rural Limerick. Apart from the rural/urban divide, people from these two areas would not usually have had much association with each other and were likely to view each other with suspicion.[21] When Honora needed to defend her identity against Martha's encroachment, she used what can be termed 'rural networks'. She called upon her neighbours and with them meted out immediate punishment, moving Martha out of the symbolically charged contested spaces and into the public view of the verandah, where Martha was judged by the group of women and physically punished for presuming to challenge her mistress's position. It is in her use of informal, feminized power networks that Honora's Irish identity is clearest. The instant justice that she and her friends delivered was outside the recognized legal framework of the colony and depended for its success on personal associations and communal recognition of the slights that had been dealt to Honora by Martha. It was also the action of someone who perceived herself to be outside the formal societal power structures and unlikely to achieve recognition or redress for her complaints through them. In these ways her actions can be understood as gendered feminine, not part of the masculine and colonial power structures of the community in general. If rural 19th-century Ireland is also understood as a colonized space, (Carroll, 2003; Cleary, 2003) then Honora's actions can be seen as drawing on her own cultural memories and experiences from a place where the only legal remedies available were those of an alien and colonizing power. So although Honora was white, European and part of the colonization of Australia, her identity was initially formed in a previously colonized place – rural Ireland.

To achieve recognition of the injustice of her plight, Martha in her turn used support structures that may be termed urban and those of the colonizers – police and courts. As a Belfast-born Catholic, she might have been expected to avoid the mechanisms of the British state, such as the police, that in Belfast were perceived as heavily biased towards Protestants. While she may have known that the local constable was likely to be an Irish Catholic and so trusted him to help her, Martha's use of the police and court system was also mediated through her employer, Robert Jenkins, Protestant Scots-born and respected citizen of Glenorchy. Robert called the doctor to attend to her and gave evidence for her in the court case. In her fight, Martha drew on resources that were gendered masculine and part of the colony's legal system. As an Irish-born Catholic woman, her use of these strategies is in contrast to Honora's instant, rural, feminized justice, and signals both the complexity and heterogeneity of Irish women's identities in colonial contexts. Even in local events such as this, some Irish born women participated and were complicit in British colonial power structures, while others resisted in their own way.

When the incident was aired in the public and masculine spaces of colonial courts and newspapers, the informal, rural actions of Honora and her friends were feminized and trivialized. Although there was evidence that five friends of Honora's had been involved in the argument, it was the four women who were fined while the case against James Blair was dismissed, even though Martha testified that James had said 'pull the _____ out and tear her guts out'. Robert

Jenkins however testified that there was 'a mob of women' attacking Martha, thus silencing James's involvement and relegating the dispute to a mere women's quarrel. In the other cases Honora and her friends, Jane Payne and Bridget Fitzgerald, were all bound over to keep the peace, in judgements that effectively vindicated Martha's stance. The magistrates clearly viewed the dispute as being between the women and silently ignored the actions of the man who participated.

Conclusion

From the late 1850s to 1863, three Irish-born women lived and worked in the Royal Mail Hotel. By mapping the fluidity of their identities and their spatial expression, their heterogeneity becomes clear. Although all three were born in Ireland, they brought with them cultural memories, experiences and class backgrounds that were quite different. Emma Ryan migrated from County Tipperary as a young woman during the Famine in Ireland, though she does not seem to have been one of the more than 4000 'famine orphans' who migrated under government schemes to Australia between 1848 and 1850 (McClaughlin, 1991). Her subsequent public identity as the wife of Robert Jenkins, prominent storekeeper and later publican in the small town of Glenorchy and employer of servants like Honora Kenealy, was undermined by her private and public use of alcohol. Emma's mental and physical decline was displayed through her use of the private spaces of her household. She alone of the women who lived in the house was described as often being in bed, being insensible with drink and unable to care for herself or her baby. Her identity shifted from respectable married woman of Glenorchy to 'a beast' through her use of alcohol. After her death her husband erected a handsome gravestone, ensuring that her identity and memory after death was that of a solid respected wife of the wealthiest man in town.

Honora Kenealy's identity shifted when she married Robert Jenkins. Although she was probably managing the household in the final years of Emma's life, she officially became mistress of the hotel rather than domestic servant on her marriage in 1861. This was not the permanent shift in social status that she might have hoped. Her relationship with Robert had obviously soured by 1863, and this is mapped in the spatial construction of the social relationships in the household, with the servant Martha keeping the keys to the cupboards and Honora sleeping with the children and nursemaid rather than in the 'adult' areas of the household. The public recognition of Honora's loss of status as a wife and mistress in the hotel was expressed in Martha's appropriation of the parlour in which she sat and sewed for one of the children. Though her duties as servant might include sewing clothes for the children, her occupation of the household spaces usually constructed as the public domain of a wife and mistress was interpreted by Honora and the women of the town as provocative and insulting. Honora used strategies that have resonances of rural Ireland when she and her friends removed Martha from the contested space of the parlour and out into the public arena of the verandah and physically punished her for her actions. Martha's networks were different from Honora's. She lived far from family and there is no evidence of any connections in Glenorchy except with her employer. The lawyer appointed to defend Honora told the court that since there had been constant bickering in the hotel, the servant should have left. She chose to stay and to resist the informal feminized strategies that Honora used to reclaim her social status. By using the urban, and masculine legal structures of the colony, Martha expressed her Irishness in different ways to Honora, emphasizing her departure from her Irish origins and her adoption of the colony as her home.

In many ways later events vindicated Martha and her uses of colonial legal remedies, suggesting that rural, feminized and localized strategies remembered from Irish townlands did not have a place in the emerging colony of Victoria. Honora lost her battle for her husband's affections. At the time he made his will in 1875, they were living separately and he left her the sum of £5 out of an estate of nearly £1000, the same amount he left to an estranged son. Although Honora was living in a property that he owned, he left the house and shop to another son. Martha however was on the first step to increased respectability, using the pathway of marriage and working in hotels familiar to many Irish migrants. She married and was widowed twice and, after her second husband's death, she ran several hotels in Stawell and at her death in 1902 left an estate worth over £2000 to her daughter who took over the hotel.[22] The report in the press of her sudden death noted the concern of her servant girl who had cared for her during the night before her death.[23] It seems that Martha had managed to negotiate the delicate relationship between mistress and servant more successfully than Honora had 40 years before.

This study of local incidents that occurred within one household over a four-year period highlights the importance of analysing the spatial expressions of Irish women's identities as they changed over life cycle and through circumstance. Women who were born in Ireland did not necessarily share similar backgrounds and experiences before arriving in Australia. Investigation of the particular geographical and cultural memories that they might have brought with them is essential to understanding how they constructed their identities as Irish women in the colonies. Irish women, such as Honora Kenealy, Martha O'Kane and perhaps Emma Ryan, usually arrived as single women within kin networks, initially worked as domestic servants and then moved on to married life that often involved employment of servants themselves. In working out the private and public expressions of servant and mistress identities, social construction of the meanings of space was crucial. Though the dispute between Honora and Martha was an extreme example, the description of the incident in the press allows us to map the contested sites of work and leisure within the Royal Mail Hotel and to view the construction of meaning that these Irish women and their friends placed on these sites. While these expressions of Irishness are not as obvious as the masculine ones shown in St Patrick's Day races and balls held every year, analysis of them reveals the heterogeneity of Irish identity that is obscured by speeches such as those of William O'Callaghan under a shamrock decorated harp.

Notes

1   The research for this chapter was funded by the Leverhulme Trust UK as part of a larger project "Memory, Place and Symbol: Irish Identities and Landscape in Colonial Australia" and headed by Lindsay Proudfoot. A version of this chapter was presented at the *International Federation for Research of the History of Women* conference, Belfast, August 2003. I would like to thank Dorothy King and Wendy Melbourne for sharing with me their knowledge of the history and people of Stawell as well as generous access to the records of the Stawell Historical Society, also Lindsay Proudfoot, Elizabeth Malcolm, Dolly McKinnon and Louise Willis for helpful comments on earlier versions of this paper.

2   *Ararat and Pleasant Creek Advertiser*, 22 March 1863.

3   *Ararat and Pleasant Creek Advertiser*, 14 July 1863.

4   *Ararat and Pleasant Creek Advertiser* 12 and 19 May 1863.

5   *Census of Victoria, 1861.* (1863-4) Population is collated by age for the electoral district of Glenorchy, which included the surrounding rural areas. There were 220 females and 616 males in the electoral district – of these there were 101 male and 101 female children under 15 years of age.

6   Eight have been positively identified as being from Ireland and a further five have Irish names. These figures have been derived from cross checking the Stawell Roads District Rate roll of 1863, VPRO VPRS 6896. With the *Pioneer Index Victoria Births Deaths and Marriages 1836-1888.* (1998) Melbourne: Macbeth Genealogical services and the Department of Justice Victoria. There are also another 21 women who registered the births of children in Glenorchy in 1861 and 1866 (1862-1865 cannot be distinguished in the BDM index). These women may have been living in Glenorchy or more likely they were living on properties in the area around Glenorchy.

7   Births are calculated from the BDM indices between 1861 and 1866.

8   Stawell Borough Council Rate books, VPRO VPRS 6896 volume 1.

9   Patrick D'Arcy's hotel licence, *Ararat and Pleasant Creek Advertiser*, 19 June 1863; land purchases, *Pleasant Creek News,* 18 September 1868, 21 August 1869, *Stawell News*, 10 August 1882 and his membership of civic committees, *Pleasant Creek News* 5 August 1869, and Ord (1896) For George Jennings see his obituary *Pleasant Creek News*, 3 September 1897.

10 *Ararat and Pleasant Creek Advertiser*, 1 July 1864.

11 Probate documents, VPRO VPRS 24/p/0000/105 18/582.

12 Biographical details are from Emma Jenkins' death certificate and Robert Jenkins' death certificate. Copies of both are held by the Stawell Historical Society.

13 Joseph Jenkins see *Stawell News 12th Nov. 1889.* For Mary Cowlands nee Jenkins see *Pleasant Creek News* 11 January 1910.

14 Shipping records, VPRO VPRS 7310, vol. 13, pp. 296 and 302.

15 *Pleasant Creek News*, 1 September 1868, 26 June 1869; (Stawell (Pleasant Creek) cemetery register and headstones 1858-1983, transcribed by Dorothy King).

16 Shipping records, VPRO VPRS 24/p/0000/105, p. 296.

17 Marriage certificates St Patrick's parish church, Stawell. Copies held by the Stawell Historical Society.

18 Inquests into Emma Ryan and Emma Jenkins' deaths: VPRO VPRS 24/p/0000/105 and 24/P/0000/39 1857/101.

19 Inquest, VPRO VPRS 24/p/0000/350 1876/103.

20 Will VPRO VPRS 7591/p/0002 unit 45 18/582.

21 In the 1930s the Belfast-born husband of Limerick woman Angela McCourt was treated with suspicion by her relatives.

22 Will VPRO VPRS 7591/p/0002/339.
23 *Pleasant Creek News,* 19 August 1902.

References

Baynton, B. (1902), *Bush studies,* Duckworth, London.
Blunt, A. and Rose, G. (1994), 'Women' Colonial and Postcolonial Geographies', in A. Blunt and G. Rose (eds), *Writing Women and Space: Colonial and Postcolonial Geographies,* The Guilford Press, New York and London, pp. 1-28.
Campbell, M. (1991), 'Irish-women in Nineteenth Century Australia: A More Hidden Ireland?' in P. Bull, C. McConville and N. McLachlan (eds), *Irish-Australian Studies: Papers Delivered at the Sixth Irish-Australian Conference,* La Trobe University, Melbourne, pp. 25-38.
Carroll, C. (2003), 'The Nation and Post-colonial Theory', in C. Carroll and P. King (eds), *Ireland and Post-colonial theory,* University of Notre-Dame Press. Notre-Dame, pp. 1-15.
Census of Victoria, 1861 (1863-4), J. Ferres, Government. Printer, Melbourne.
Clarke, M. (1976), 'An Up-country Township', in M. Wilding (ed.), *Portable Australian Authors: Marcus Clarke,* University of Queensland Press, St Lucia, pp. 570-75 (first published 1870).
Cleary, J. (2003), 'Misplaced Ideas? Colonialism, Location and Dislocation in Irish Studies', in C. Connolly (ed.), *Theorising Ireland: Readers in Cultural Criticism,* Palgrave Macmillan, Basingstoke, Hampshire, pp. 91-104.
Dingle, A.E. (1980), '"The Truely Magnificent Thirst": An Historical Survey of Australian Drinking Habits', *Historical Studies,* vol. 19, pp. 227-49.
Fitzpatrick, D. (1994), *Oceans of Consolation: Personal Accounts of Irish Migration to Australia,* Cork University Press, Cork.
Fitzpatrick, D. (1998), '"This is the Place that Foolish Girls are Knowing": Reading the Letters of Emigrant Irish Women in Colonial Australia', in T. McClaughlin (ed.), *Irish Women in Colonial Australia,* Allen and Unwin, Sydney, pp. 163-81.
Gothard, J. (2001), *Blue China: Single Female Migration to Colonial Australia.* Melbourne University Press, Melbourne.
Gothard, J. (2001), 'Wives or Workers? Single British Female Migration to Colonial Australia', in P. Sharpe (ed.), *Women, Gender and Labour Migration: Historical and Global Perspectives,* Routledge, Florence, KY, pp. 145-62.
Haines, R.F. (1996), 'Workhouse to Gangplank: the Mobilization of Irish Women and Girls Bound for Australia c1850', in R. Davis, J. Livett, A.-M. Whitaker and P. Moore, *Irish-Australian Studies: Papers Delivered at the Eighth Irish-Australian Conference, Hobart July 1995,* Crossing Press. Sydney, pp. 166-76.
Haines, R.F., Kleinig, M., Oxley, D., and Richards, E. (1998), 'Migration and Opportunity: An Antipodean Perspective', *International Review of Social History,* vol. 43, pp. 235-63.
Hamilton, P. (1993), 'Domestic Dilemmas: Representations of Servants and Employers in the Popular Press', in S. Magarey, S. Rowley and S. Sheridan (eds), *Debutante Nation: Feminism Contests the 1890s,* Allen and Unwin, St. Leonards, NSW, pp. 71-90.
Jackson, P. (1984), 'Women in Nineteenth Century Irish Emigration', *International Migration Review,* vol. 18, pp. 1004-1020.
Jacobs, J. and Nash, C. (2003), 'Too Little, Too Much: Cultural Feminist Geographies', *Gender, Place and Culture,* vol. 10, pp. 265-79.

Johnson, L., Huggins, J. and Jacobs, J. (2000), *Placebound: Australian Feminist Geographies, Meridian: Australian Geographical Perspectives,* Oxford University Press, Melbourne.

Kingston, R. (1989), *Good Country for a Grant: A History of the Stawell Shire,* Shire of Stawell, Stawell.

Lawrence, R.J. (1982) 'Domestic Space and Society: A Cross-Cultural Study', *Comparative Studies in Society and History,* vol. 24, pp. 104-130.

Luddy, M. (2000), Women and Work in Nineteenth and Early Twentieth Century Ireland, An Overview, in B. Whelan, *Women and Paid Work in Ireland 1500-1930,* Four Courts, Dublin, pp. 44-56.

Massey, D.B. (1994), *Space, Place and Gender,* Polity, Cambridge.

McClaughlin, T. (ed.), (1991), *Barefoot and Pregnant? Irish Famine Orphans in Australia: Documents and Register,* Genealogical Society of Victoria, Melbourne.

McClaughlin, T. (1996), 'Vulnerable Irish Women in Mid-Late Nineteenth Century Australia', in R. Davis, J. Livett, A.-M. Whitaker and P. Moore (eds), *Irish-Australian Studies: Papers Delivered at the Eighth Irish-Australian Conference Hobart July 1995,* Crossing Press, Sydney, pp. 157-65.

McClaughlin, T. (ed.), (1998), *Irish Women in Colonial Australia,* Allen and Unwin, Sydney.

McCourt, F. (1996), *Angela's Ashes: A Memoir,* Harper Collins, London.

McDowell, L. (1992), 'Doing Gender: Feminism, Feminists and Research Methods in Human Geography', *Transactions of the Institute of British Geographers,* vol. 17, pp. 399-416.

McPherson, B. (1994) 'The Verandah as a feminine site in the Australian memory', in J. Hoorn (ed.), *Strange Women: Essays in Art and Gender,* Melbourne University Press, Melbourne, pp. 67-80.

Mills, S. (1996), 'Gender and Colonial Space', *Gender, Place and Culture,* vol. 3, pp. 125-47.

Morgan, S. (1989), 'Irishwomen in Port Philip and Victoria 1840-1860', in O. MacDonagh and W.F. Mandle (eds), *Irish-Australian Studies Papers Delivered at the Fifth Irish-Australian Conference,* Australian University Press, Canberra, pp. 231-49.

Morin, K.M. and Berg, L.D. (1999), 'Emplacing Current Trends in Feminist Historical Geography', *Gender, Place and Culture,* vol. 6, pp. 311-330.

Ord, M. (1896), *Stawell Past and Present,* Stawell News and Pleasant Creek Chronicle, Stawell.

Payton, P. (2001), 'Cousin Jacks and Ancient Britons: Cornish Immigrants and Ethnic Identity', *Journal of Australian Studies,* vol. 69, pp. 54-64.

Pickering, M. (2002), 'Looking Through the Fawn-skin Window: White Women's Sense of Place in the New Worlds of Australia and Canada', *Australian Historical Studies,* vol. 118, pp. 223-38.

Probyn, E. (2003), 'The Spatial Imperiative of Subjectivity', in K. Anderson, M. Domosh, S. Pile and N. Thrift (eds), *Handbook of Cultural Geography,* Sage, London, pp. 290-99.

Rowley, S. (1994), 'The Journey's End: Women's Mobility and Confinement', in J. Hoorn (ed.), *Strange Women: Essays in Art and Gender,* Melbourne University Press, Melbourne, pp. 81-97.

Rule, P. (1996), '"Very Lonesome...in a Foreign Land" Female Reconstruction of Irish Families in Colonial Victoria', in R. Davis, J. Livett, A.-M. Whitaker and P. Moore (eds), *Irish-Australian Studies: Papers Delivered at the Eighth Irish-Australian Conference,* Crossings Press, Sydney, pp. 179-88.

Rule, P. (1998), '"Tell Father and Mother not to be unhappy for I am very comfortable" a Sketch of Irish Women's Experiences in Colonial Victoria', in T. McClaughlin (ed.) *Irish Women in Colonial Australia*, Allen and Unwin, Sydney, pp. 123-41.

Stawell (Pleasant Creek) cemetery register and headstones 1858-1983, transcribed by Dorothy King, *Stawell Historical Society*, Stawell.

Wallace, J. (1888), 'Glenorchy and Wimmera Chronicle', *Stawell Times*, 9, 19, 30 March; 7, 21 May; 20 June; 12 November; 17 December.

Walter, B. (2001), *Outsiders Inside: Whiteness, Place and Irish Women*, Routledge, London and New York.

Warsh, C.K. (1993), '"Oh Lord, pour a corkial in her wounded heart." The Drinking Woman in Victorian and Edwardian Canada', in C.K. Warsh (ed.), *Drink in Canada. Historical Essays*, McGill-Queens University Press, Montreal and Kingston, pp. 70-92.

Watts, R. (2003), Making Numbers Count: The Birth of the Census and Racial Government in Victoria 1835-1840, *Australian Historical Studies*, vol. 121, pp. 26-47.

Wright, C. (2001), Of Public Houses and Private Lives: Female Hotelkeepers as Domestic Entrepeneurs, *Australian Historical Studies*, vol. 116, pp. 57-75.

Wright, C. (2003), *Beyond the Ladies Lounge: Australia's Female Publicans*, Melbourne University Press, Melbourne.

Chapter 6

Performing Power, Demonstrating Resistance: Interpreting Queen Victoria's Visit to Dublin in 1900

Yvonne Whelan

Introduction

In September 1998 while on an unofficial visit to London, the Irish President, Mary McAleese announced her intention to join with Queen Elizabeth II at Messine, Belgium for the opening of the British-Irish funded memorial in honour of the Irish service men who died during World War II. At the same time she also confirmed that a meeting with the Duke of Edinburgh in Dublin had been scheduled for later in the year. Both developments were hailed by the President as a sign of the leadership being given in creating 'fully grown, fully adult relationships between these two islands [of Britain and Ireland]', (*The Irish Times*, 18 September 1998). Moreover, McAleese looked forward to a time when she could welcome the British monarch on a state visit to Dublin, 'I think the day that happens is a day we can all say 'Yes, we've arrived at a degree of comfort with each other that really does mark the closing of the culture of conflict... it will mark the budding-in of the culture of consensus that we are building' (*The Irish Times*, 18 September 1998). After half a century on the throne, such a visit by the monarch in the closing years of her reign would be an enormously significant event. It would also mark the first visit by a British monarch to Ireland since that of King George V in 1911 and the first since the achievement of political independence in 1922.

Less than a hundred years ago Dublin was the deposed capital of a country that endured a strained and ambivalent colonial relationship with its nearest neighbour, Great Britain. Successive rebellions during previous centuries had failed to secure political independence but as a new century beckoned, constitutional nationalists strove to secure a measure of legislative independence by peaceful means and with that the Home Rule movement gathered pace. A series of Home Rule Bills were put before Parliament but the intervention of World War I put a temporary stop to political progress. It was, however, amid this turbulent political context that a series of visits to Ireland by British monarchs took place and during these events both urban and rural landscapes throughout the country served as stages upon which imperial power was acted out with a large measure of pomp and ceremony.

The first 11 years of the century saw three monarchs make five visits; Queen Victoria (1900), King Edward VII (1903, 1904 and 1907) and finally King George V (1911). These visits were invariably politically motivated, well orchestrated and their timing carefully planned. Cities and towns throughout the country were lavishly decorated, royal processions through urban spaces were symbolically planned, gigantic temporary structures erected and addresses of welcome carefully crafted by local councils and municipal corporations. The cultural landscape was effectively transformed and space became politicized in a manner that underscored Ireland's status as partner in the British imperial enterprise. In Dublin, they served to cultivate a landscape of imperial power in what was then a city of contested space. The parades that accompanied each visit, the routes that were followed through the city, the display and theatricality that went with them, along with the composition and involvement of parade participants and the symbolic devices employed throughout, each ensured that these landscape spectacles were highly successful in constructing and sustaining a sense of imperial identity. They also provoked enormous public interest as well as rivalry between various groups vying to express their loyalty to the crown.

This, however, is only one reading of what was in fact a much more complex and fraught political relationship that prevailed at this 'contact zone' between Britain and Ireland. Each visit actually engendered ever-increasing degrees of opposition and fervent resistance, which threw into sharp relief the highly contested status of Ireland as imperial partner. Various strands of Irish nationalist, republican and socialist opinion used the royal visits in order to galvanize opposition to the empire and to demonstrate that Ireland, far from being an imperial partner, laboured under a malign form of colonial rule from which every effort should be made to break free (Paseta, 1999). In this chapter I will focus on the contentious debate that surrounded the visit of Queen Victoria to Dublin in April 1900 and the royal procession through the centre of Dublin city that marked the onset of her three-week stay in Ireland. At the outset I will briefly consider Ireland's contested status as imperial partner or colonial subordinate at the turn of the 20th century. This was by no means clear cut, indeed, the extent to which we can refer to Ireland as being either colonial or imperial has been the subject of much debate (Kenny, 2004; Kennedy, 1996 and Bartlett, 1988). I then focus on the role of public spectacles and processions as important sources to be utilized by geographers in unravelling the geographies of hegemony and resistance. These conceptual issues are then explored in greater detail with reference to the spectacle of the Queen's 1900 visit. Particular attention is paid to the role of this event in underscoring Ireland's status as imperial partner and to the ways in which the urban landscape of the city of Dublin was used in order to underpin this narrative. This necessitates an examination of the ritual and well-choreographed spectacle that marked the visit, including the use of decorations, colour, the erection of temporary structures, the presentation of an address of welcome, as well the use of public space for the purposes of royal procession through the city on the day of the Queen's arrival. Finally, the emphasis shifts to an exploration of alternative readings of these events and to the ways in which they were actively resisted by the growing ranks of Irish republicans, nationalists and socialists.

Dublin: A Contested Capital City

Amidst the political discontinuities and tensions of the early 20th century, Ireland's colonial status was far from agreed. A mid-latitude colony of settlement rather than a tropical colony of exploitation, the settler population in Ireland was small and the colonial experience was different to that experienced elsewhere and especially in the non-western world. The Head of State was the British monarch, who was represented in the country by his viceroy, while power rested with the Chief Secretary for Ireland, upon whose advice the viceroy acted. As Fitzpatrick argues, 'both in form and in practice, the government of Ireland was a bizarre blend of "metropolitan" and "colonial" elements. On the one hand a complicit partner in Britain's imperial exploitation, Ireland was also an internally colonized subject, although unlike Britain's transoceanic colonies, the island enjoyed a substantial measure of representation in the parliament at Westminster and was in the eyes of many integral to the composition of the United Kingdom' (Fitzpatrick, 1998, p. 6). Ireland's status as a colony, therefore, was rather ambivalent and significantly different from, for example, India and Calcutta, its counterpart capital and vice-regal seat of the British Indian Empire. As Stephen Howe suggests of Ireland at the turn of the century, it was 'a sphere of ambiguity, tension, transition, hybridity, between national and imperial spheres', in a manner that was by no means unique but was echoed in places like Canada or Cardiff, Auckland or Aberdeen (Howe, 2000, p. 68). Ireland, therefore, occupied a rather ambiguous position within the empire, being a contested terrain and a space of hybridity between the imperial centre and the margins. As Gibbons argues, 'Ireland is a first world country, but with a third world memory. Though largely white, Anglophone and westernized, Ireland historically was in the paradoxical position of being a colony within Europe', a fact which makes it fertile territory for post-colonial studies (Gibbons, 1998, p. 27).

Queen Victoria's visit to Ireland at the turn of the 20th century threw into sharp relief the country's contested colonial status. On 1 January 1900 *The Irish Times* editorial wrote of the 'marvellous growth of the National prosperity'. The nation that the author was referring to was that of the United Kingdom of Britain and Ireland and the article went on to point out that 'the progress of the United Kingdom has been of a steady and sterling character and never at any time in the course of all British history, have Imperial interests been founded upon so strong and solid a basis'. Ireland was portrayed as part of perhaps the greatest empire the world had ever seen and it seemed clear that whatever change the 20th century might bring to the country, it would take place within the context of the 'steady and sterling progress' of the British Empire. However, there also existed a significant proportion of the country's population that saw Ireland's future fortunes in an altogether different light and which rejected its position as constituent element of the empire. This loose grouping aspired towards Irish autonomy and Home Rule within the United Kingdom, the struggle for which had preoccupied nationalist politicians since the 18th century. Alongside these constitutional Home Rulers, there was a more militant fringe of nationalists who operated outside the bounds of the Irish party at Westminster and whose goal was a completely independent Irish republic, to be achieved by whatever means necessary. In 1900 this group expressed itself most forcefully with demonstrations of support for the Boers in their war against the British. As Paseta (1999) points out, in the early weeks of the Boer War the Irish Transvaal Committee, led by members of the Dublin intelligentsia, mounted an outdoor screen in College Green, Dublin, onto which they projected text and photographs hailing the Boers' success against the British. Each of these cohorts, of agencies loyal to empire, of Home Rule nationalists and of more militant republicans, collided in 1900 during the visit of Queen Victoria, and especially during the spectacle that marked the Queen's arrival in the city. Before exploring in more detail these aspects of the royal visits, however, I want to elaborate upon the role of parading and performance in articulating and underpinning particular narratives of identity.

Parading Power and Resistance in the Cultural Landscape

In interrogating the 'real geographies' of colonialism, imperialism, and postcolonialism, aspects of the cultural landscape have become important tools of geographical inquiry. In recent years, geographers have explored the ways in which public statuary, street names, architecture and urban design serve as significant sources for unravelling the geographies of political and cultural identity. It is often argued that the dynamic relationship between history and geography is demonstrated when, for example, national monuments, public buildings and streets celebrating national heritage are inserted into the landscape in a manner that maps history onto territory. These landscape elements act as spatializations of power and memory, making tangible specific narratives of nationhood and reducing otherwise fluid histories into sanitized, concretized myths that anchor the projection of national identity onto physical territory (Bell, 1999). As Kong, (1993, p. 24) argues, 'landscapes are ideological in that they can be used to legitimize and/or challenge social and political control'. In this chapter, I pay particular attention to the role of procession, parading and public performances in the arena of the urban landscape in order to trace the expressions of both subaltern resistance and hegemonic power. Endowed with meaning on a number of levels, public parades not only serve as rituals of remembrance, but they are also carefully considered expressions of power. As Jarman and Bryan (1998, p. 1) point out, 'Parades are not simply cultural asides, elements of a tradition which reveal the historical roots of a community, rather they have been and remain, pivotal in defining the relationships between the state and local communities. The history of parades is the history of community relations and of power relationships'.

In the hands of hegemonic powers, therefore, parades take on a particular significance. As Marston observes, parades are 'complex commentaries' on social relations, they are both shaped by the field of power relations in which they take place and are attempts to act on and influence those relations (Marston, 1989; see also Davis, 1986). In their work on ritual and spectacle in post-independent Singapore, for example, Yeoh and Kong observe that 'national day' parades played an important role in inventing the imagined community of the nation, 'parading, by traversing and transforming ordinary spaces, allows the symbolic capture of spectacles to move beyond the locus of the ceremonial landscape to the habitations of the people' (Yeoh and Kong, 1997, p. 222). A burgeoning body of work has highlighted the significant role of parades as a form of identity expression and demonstrates that as landscape metaphors, parades have an impact both in terms of their actual substance, as well as through pageantry, fanfare and show, through the demonstration of military might, through the deliberate use of colour, through the manipulation of lights and fireworks, and through the orchestration of sound and music (Yeoh and Kong, 1997, p. 224. See also Fraser, 2000; Goheen, 1990, 1993, 1994, 1998; Jarman and Bryan, 1998; Marston, 1989). There are, therefore, a range of attributes associated with parading and public spectacle which must be considered. These aspects include; the ceremonial space that is the parade route, the different aspects of display and theatricality that go hand in hand with the public performance, including the use of decorations, temporary structures, lighting, fireworks, music, colour, as well as the role of the military.

In exploring the spectacle of public performance that accompanied royal visits to Ireland, we will see, however, that not only did such events serve to buttress popular support for the imperial power, but they also served as focal points around which nationalist resistance to such hegemony found expression. While dominant regimes use history, heritage, memory and various landscape elements in order to memorialize specific narratives of identity, such narratives can also be challenged and resisted by those less powerful, especially in colonial contexts. Consequently, in contested political contexts where some groups do not identify with the dominant elite but rather seek empowerment for themselves, the cultural landscape often becomes a site of conflict and contestation. While traditionally geographers have tended to focus upon the historical geography of colonialism with the emphasis firmly fixed on the processes and patterns of colonial domination, this has been destabilized in recent years in an attempt to accommodate what Yeoh refers to as a 'politics of space in the colonized world where people resisted, responded to and were affected by colonization' (2000, p. 146). Hence, geographers are now more sensitive to the complexities and multiplicity of resistances that prevail at the contact zone between the colonizer and the colonized. Consequently, within cultural geography there has been a repositioning from an engagement purely with geography and empire where the emphasis is firmly focused on the colonizing power, to a consideration of the multiple historical geographies of the colonized world. Hence, landscape is now read as a site of conflict and collision, negotiation and dialogue between the colonizer and the colonized. Moreover, those sites, symbols and spectacles of dominance which make concrete the power of prevailing regimes, also serve as focal points around which the marginalized and less powerful can orchestrate opposition and resistance.

Performing Imperial Power: The Visit of Queen Victoria

Each of the royal visits to Ireland in the early years of the 20th century embodied a significant measure of spectacle, theatricality and choreographed ritual, and was designed to ensure a measure of symbolic and indeed political significance that went far beyond the remit of a merely private visit. As spectacles, they temporarily transformed cities and towns, and in creating a sense of awe and wonderment they served to cultivate a narrative of imperial identity that brought Irish people into a closer communion with the centre of the empire. They also, however, met with much resistance and instigated fierce debate within the nationalist community who challenged the imperial hegemony that these events represented. This was clearly demonstrated not only *during* the visit of Queen Victoria in 1900, but also in the weeks leading up to the visit when the city was being suitably prepared for the arrival of the monarch and a heated debate broke out regarding the proposal to present an address of welcome.

In early 1900 an announcement came from Dublin Castle, the seat of the British administration in Ireland, that Queen Victoria was to visit Ireland. The announcement came shortly after the Queen had authorized that Irish soldiers serving in South Africa would be permitted to wear a shamrock on St Patrick's Day. Although Castle officials stressed the informal and private nature of the visit, suggesting that the Queen merely desired a 'change of air', she was to use her time in Ireland to undertake a number of official engagements, which left many nationalists suspicious of the supposedly private nature of the visit. For example, the Queen commended the action of the Royal Dublin Fusiliers for their part in the Boer War and encouraged conscription to the cause. She also took part in two other public ceremonies; the review of the Dublin Garrison in the grounds of the Viceregal Lodge and an inspection of school children during Children's day in the Phoenix Park on the first Saturday of her visit (Condon, 2000). During the three weeks of her stay Ireland was swept up in a tide of imperial enthusiasm and it was the royal procession through the city on the day of her arrival that marked the first, and perhaps most significant aspect of the imperial spectacle. One of the first tasks facing state officials before the Queen ever arrived in Dublin, however, was to prepare the city for her arrival and give it a suitably imperial gloss which would in turn provide citizens with vivid images and help foster a sense of imperial identity and citizenship, while also educating them in the ideals of imperialism.

Decorating Dublin

In the weeks leading up to the royal visit no expense was spared in decorating Dublin. Local councils throughout the city established committees and efforts were especially fervent along the route of the royal procession. *The Times* newspaper claimed that the decorations would, when finished, 'produce a display of loyalty absolutely unparalleled in the records of public rejoicing' (*The Times*, 5 April 1900). Other newspapers anticipated that:

The period of the Queen's sojourn will be a brilliant one. Irish loyalty will find its full and just representation ... From the moment that the Royal procession starts from Kingstown until the Viceregal Lodge is reached it will pass between groves and garlands of flags and evergreens, and under not a few triumphal arches. But a nobler tribute will still be offered to Her Majesty by the presence of her subjects who will crowd the many stands and line the roads, tendering to her a hearty and respectful greeting, one which comes from their hearts, which is unaffected, and which is sincere. The Queen's Day in Dublin will be memorable in the annals of our time and we look to it as the opening of an era of a larger-hearted kindliness, and of a better spirit of friendliness amongst the citizens of the British Empire (*The Irish Times*, 4 April 1900).

The Queen's procession through the city was to take her from Kingstown in the south to the city centre where she was to be presented with the keys to the city by members of Dublin Corporation, before proceeding to the Viceregal Lodge in the Phoenix Park on the city's north side. In the days leading up to her arrival the image of the city was effectively transformed as streets all along the route were bedecked with bunting, Venetian masts, triumphal arches, Union Jacks and Royal Standards. Giant floral displays were erected. Illuminated words of welcome sprang to life at night, harnessing the power of the night sky. Even streets off the main parade route were decorated with striking use made of appropriate colours, especially crimson and royal blue, 'almost every house from this place forward is decked with bunting, and in many cases the windowsills and balconies are draped with crimson and gold or royal blue and silver dressings' (*The Irish Times*, 2 April 1900). A correspondent in *The Irish Times* wrote that:

The decoration of the city goes on apace, and now both the general public scheme and the displays of individual establishments have reached majestic proportions ... gigantic operations have been carried on, and most of the houses ... are looking bright and radiant in their rejuvenated guise. Then the displays of the various emblems of welcome are superb in their richness, scope and variety ... The route of the Royal progress is naturally the chief element in this wonderful display. A striking manifestation of unanimous and enthusiastic welcome is presented from beginning to end of the lengthy drive ... Not only have existing places of natural and architectural beauty been greatly enhanced by brilliancy of colour, but the very bare, dilapidated, and even repulsive spots along the route have undergone an extraordinary reformation, and offered fresh fields for the exercise of decorative fancy (2 April 1900).

Contemporary newspapers carried detailed commentaries on the work of the decoration committees that had been appointed throughout the city and recounted in detail their decorative endeavours. At Upper Leeson Street, for example, which was to act as one of chief focal points of the Queen's visit, a mock medieval gate was erected where the Queen was to perform the medieval custom of formally demanding admission to the city, before being presented with the City Keys, the Sword of State, and the Corporation Address. Further along the royal route bunting lined public thoroughfares, while both public and private houses were lavishly decorated. At College Green in the heart of the city, both Trinity College and the old Parliament House were extensively decorated. At the college, 'poles swathed in crimson baize are topped by harps, and each of the electric tram poles is similarly set off by a gilt crown', while 'the decorations of the Old Parliament House also stand out prominently in a scene of brilliant decorative triumphs. Across the central columns of this stately edifice the word "Welcome" is set out in illuminants, and coloured lights wind spirally up the columns on either side' (*The Irish Times*, 2 April 1900). Ultimately, 'decorations of one kind or another meet the eye at every turn, and amongst the people in the streets badges of emblematic loyalty to the Crown are worn to an extent never witnessed on any previous occasion ... Dublin has prepared for Her Majesty's entry, and has been adorned as never it has been before' (*The Irish Times*, 4 April 1900).

These decorative endeavours were much more than an attempt to prettify the city, however. Rather, they played an important role in creating the imperial spectacle and they were to leave an indelible impression upon onlookers. It is important to note, however, that not all were content the work of the various committees. In fact there was a great deal of friction in the city and opposition was voiced to the ways in which the streets of Dublin were being decorated, barely masking a more deep-seated antagonism towards the Queen's visit to Ireland. One commentator in the nationalist *Irish People* newspaper, for example, wrote of:

> Long gaudy lines of colours put up somehow and anyhow – and all foreign... they do not suggest an Irish welcome. They only indicate that wealthy Dublin Unionists have... given the Southern side of Dublin the appearance of an English city for the time being (*The Irish People*, 7 April 1900).

The article went on to complain that:

> Bunting from everywhere that an English flag could be spared polluted the air and desecrated buildings ... the sight of Dublin gaudily decked out in anticipation of the visit of an English sovereign is sad and nauseating ... Every 'Union Jack'... is an outward and visible token of the degradation of the capital – a symbol of the still dominant power of the English ascendancy ... Under all the banners and floral devices and glaring illuminations, was still poor old humdrum Dublin, with its 200,000 dwellers in squalid tenements, its ruined trade, its teeming workhouses, its hopeless poverty – the once-proud capital of a free nation degraded to the level of the biggest city of a decaying and fettered province (*The Irish People*, 7 April 1900).

The antipathy that prevailed in nationalist quarters regarding the transformation of Dublin signalled the tensions that existed in the city over the impending visit. In fact, it could be argued that the preparations for the parade were as hotly contested as the parade itself and not only regarding the decoration of the city, but also in relation to the presentation of an address of welcome. It is also noteworthy that this contestation existed not just between the nationalists who opposed the visit and the imperialists who went out of their way to present a loyal welcome to the Queen, but also *among* Irish nationalists. In the weeks leading up to the visit debate raged in the city about whether or not an address of welcome should be presented to the visiting monarch. More militant nationalists argued that although the visit was flagged as 'private', it was in fact aimed at luring young Irishmen to South Africa to fight with the British against the Boers, and they took great exception to the proposal by some members of Dublin Corporation to present an address of welcome to the Queen on behalf of the municipal authority (*The Irish People*, 10 March 1900). Feelings ran especially high as the Lord Mayor, Thomas Pile, who proposed the motion, had been elected as a nationalist member (*The Freeman's Journal*, 3 April 1900). Other nationalists, however, were in favour of the address and argued that it contained no single expression of loyalty to the British connection and that it was merely a personal welcome. By a margin of 32 to 22 votes the authority passed the motion to present the address. This prompted a hostile reaction from some, who argued that the mere fact of an address 'was the signal for offensive Unionist exultation and determined Nationalist protest'. They later organized a counter meeting at which they condemned the presentation of an address on behalf of all patriots. In *The Irish People* attention was drawn to the 'traitors to Ireland' who had voted for the address and it was suggested that the Lord Mayor was 'a disgusting creature', who was: 'getting ready to throw himself at her Majesty's feet with the slavish address of welcome and loyalty...' (7 April 1900).

At the next meeting of Dublin Corporation, however, a resolution was passed which used the debate over the address of welcome in order to express broader objections to British rule in Ireland. Councillors voted to adopt the following resolution:

> That, inasmuch as a section of the Unionist Press in Great Britain and Ireland have interpreted the voice of the Council to present an address of welcome to her Majesty as an abandonment of our claim for National Self-Government, this Council; assembled on the centenary of the passing of the Act of Union, hereby declares that this Act was obtained by fraud and shameful corruption; that the people of this country can never give to the system of government so established their loyal support, and that, so far as the vast body of the people are concerned, there will be neither contentment nor loyalty in this country until our National Parliament is restored (*The Freeman's Journal*, 3 April 1900).

The level of heated debate that was generated by the Queen's visit, as well as the controversy that surrounded both the decoration of the city and the presentation of the address of welcome, ensured that there was a heavy security presence in Dublin when the Queen and her entourage arrived in Dublin on the evening of Tuesday, 3 April 1900.

The royal yacht docked at Victoria pier in Kingstown, so-named after the visit of George IV in 1821 and now known as Dun Laoghaire, amidst a display of fireworks and illuminations. Newspapers recorded that the experience at Kingstown 'will scarcely be equalled within the coming century', and wrote of the 'magnificence of a loyal and almost spontaneous demonstration on the part of thousands of devoted subjects ... an enthusiastic welcome was offered to the venerable sovereign of the British empire' (*The Irish Times*, 4 April 1900). On the evening of the arrival the royal fleet was illuminated and crowds watched as:

> One after another dazzling devices were lit up by electricity and gas and sprang into notice on the principal buildings so that the town was ablaze with multi-coloured lights

... An astonishing wave of loyalty and love seemed to spread over the thousands of spectators who filled the east pier and the roads adjoining, and many persons fluttered mini-union jacks and sang patriotic airs (*The Irish Times*, 4 April 1900).

This, however, was just the prelude to the main event, the royal procession through Dublin from Kingstown to the Viceregal Lodge in the Phoenix Park which took place the following day.

'The Spectacle was a Magnificent One': Power, Resistance and the Queen's Day in Dublin

The royal party disembarked the next day to a rapturous welcome from the assembled multitude waiting on stands and lining the streets around Victoria wharf in Kingstown. The day had been declared a public holiday and following a formal address of welcome from the Kingstown Town Clerk, in which he pledged loyalty to the monarchy, the procession through the city began. The royal cortege was made up of four carriages carrying various members of the royal party, the Viceroy and his wife, and an array of government figures, while various military contingents were positioned at key points along the royal route. Although the Queen could have entered the city via the railway, 'she preferred to give the Irish people a larger opportunity for the display of that loyalty which has always been one of our strongest and more attractive characteristics' (*The Irish Times*, 5 April 1900).

The journey from Kingstown into the city was carefully orchestrated in a manner which reinforces Kuper's argument that, 'the choice of the landscape in which to stage a spectacle is not a matter of indifference, for some sites have more significance that others ... Parades do not simply occupy central space, but also move through space as a means of diffusing the effects of the spectacle' (Kuper, cited in Goheen, 1993, p. 331). The procession passed from Kingstown through Dublin's more affluent southern suburbs, where many of the prevailing street names were redolent of the city's imperial status. Drawing on the symbolic capital of the cultural landscape, the procession wended its way through the streets of the city in full view of the gathered multitude who waved and cheered from every possible vantage point. We are told that, 'every inch of space was eagerly appropriated, and every point of vantage hastily secured. Every window in every house, and in the several public offices and banking establishments, which are so thickly clustered – the National Bank, the Belfast Bank, and many of our principal insurance offices – was occupied' (*The Irish Times*, 5 April 1900). In transforming ordinary streets into theatres of pomp, the procession allowed for the symbolic capture of the urban landscape and ensured that the citizens of the city were afforded the opportunity to make a significant contribution to the spectacle. The procession was led and followed at the rear by a fleet of various military units who played a significant role in the day's events, along with several military bands which were strategically located along the route in order to provide appropriate musical accompaniment to the Queen's arrival in the city.

The royal cortege passed from the suburbs into the city centre at Leeson Street Bridge where the Queen was formally welcomed into the city and presented with the keys and civic sword. The large mock medieval castle gate and tower erected at the bridge echoed the more ancient gate to the walled city of Dublin. The 70 foot high tower was made of wood, covered in canvas and painted to imitate the 16th century stonework of the medieval Baggotrath Castle. A number of beefeaters in their traditional scarlet costumes flanked the gate and a stand was erected nearby to seat various dignitaries, among them the Lord Mayor of Dublin and other members of Dublin Corporation who had gathered to present the highly contentious address of welcome to the Queen. Meanwhile, the ancient keys of the city were placed in a golden casket upon a green poplin cushion beside the address of welcome. The set of keys were the same as those presented to the Queen during her 1849 visit, 'there are a dozen of them, each about 10 inches long, on a massive iron ring of about 5 inches in diameter. They are stamped NSW and E respectively, doubtless referring to the different gates of the city... The ceremony of presenting the keys of the city to the Sovereign is a quaint relic of bygone times and one of the ancient privileges of the city of Dublin' (*The Freeman's Journal*, 6 April 1900). After a spirited fanfare from a bugler, the ceremonial got underway when the Athlone Pursuivant of Arms, positioned outside the gate, knocked on the city gate to demand admission for Queen Victoria. In reply, the Lord Mayor directed the city gates to be thrown open, and the Athlone Pursuivant of Arms rode in, 'in all the glory of his tabard and brilliant accoutrements, the gathering became hushed, in anticipation of his message. 'My Lord Mayor of Dublin, I seek admission to the city of Dublin for Her most gracious Majesty the Queen'. The Lord Mayor rose at once, and replied in clear and distinct tones – 'On behalf of the city of Dublin, I desire to tender to the Queen a most hearty welcome to this, Her Majesty's ancient city, and·on the arrival of Her Majesty the city gates shall be thrown open on the instant' (*The Irish Times*, 5 April 1900).

With that the royal cortege then passed through the gates and the controversial address of welcome was finally presented. When her carriage stopped the Lord Mayor presented the Queen with the keys of the city, stating, 'May I tender to Your Majesty the keys of the ancient city of Dublin, and wish you a very pleasant time amongst us', (*The Irish Times*, 5 April 1900). She was then presented with the ancient sword of the city after which the Lord Mayor then called upon the Town Clerk to read the Corporation Address. The royal cortege then entered city boundary, continuing through the south inner city, passing along Merrion Square to College Green, where it received a resounding welcome from the assembled students and fellows of Trinity College:

the scene in College Green was one that will linger long and pleasantly in the memory of all who were fortunate enough to witness it. College Green, the heart of the city, and which has from time immemorial been the central point of all processions and pageants occurring within civic bounds, was yesterday the great central rallying point in the triumphal progress of the Queen, from the moment of her entry at the extemporised city gate at Leeson Street until her arrival at the Phoenix Park ... The immense open space, the magnificent architectural surroundings – Trinity College and the Bank of Ireland,

and the other splendid buildings, most of them of recent date, all helped to enhance the picturesque character of the scene. College Green probably never looked better than it did yesterday (*The Irish Times*, 5 April 1900).

The procession then moved north via Dame Street and Parliament Street where it crossed the River Liffey and made its way along the quays to the Phoenix Park and the residence of the Viceroy. It took some two hours and 22 minutes to accomplish the nine mile journey from Kingstown and it is significant that the north side of the city, which could be regarded as the more nationalist, slum ridden and poverty-stricken part, did not feature highly on the royal route. In fact, the city's main thoroughfare, O'Connell/Sackville Street, although decorated, was by-passed as the cortege travelled directly to Phoenix Park. That night the city was illuminated, 'and when darkness had fully set in a great line of light stretched along the public buildings, and banks and almost all the establishments in the heart of the metropolis. From Rutland Square in the north to Rathmines in the south, and from Merrion Square in the east, to the Park Gate in the west, was lighted up with brilliantly illuminated devices in honour of the Queen and in token of her presence in the capital. Mottoes of welcome sparkled on the house fronts, the Royal monogram glistened on all sides, the Royal Arms burned brightly on many house fronts and the pillars and pedestals and plinths of the banks were fringed with multi-coloured bulbs' (*The Irish Times*, 5 April 1900).

Contemporary newspapers commented on the 'warmth of the Queen's reception in Dublin', stating that, 'Her Majesty had a genuine Irish welcome, and her stay at the Viceregal Lodge has opened under the happiest auspices' (*The Freeman's Journal*, 6 April 1900). Reference was, however, made to the fact that the welcome afforded the Queen was not complete and that some objected to the visit on the grounds that 'the sole object of the Royal Visit was to bamboozle a few more Irish soldiers into fighting for the British government in South Africa. An Irish minor poet [Yeats] exhorted his fellow countrymen not to join in welcoming the Queen, who represented an Empire that was robbing the Boers of their liberty or if they thronged the streets to maintain a significant silence' (*The Freeman's Journal*, April 6, 1900). From the very outset nationalist objections to the royal visit were loudly declared and prominent members of Dublin-based groups set about organizing a programme of opposing events. By way of sharp contrast to the jubilation expressed in the columns of *The Irish Times*, the nationalist press was at pains to point out that the welcome received by the Queen was less than representative of the city as a whole. On the contrary, it was suggested that:

There is nothing to welcome Victoria for of course, but every reason to detest her, and therefore to leave her severely alone from the day she sets foot in Ireland until the hour she departs – to leave her to her Dublin Castle landlord and Belfast Orange Lodge friends and admirers, the only people in Ireland she ever did or does care a straw about. Leave Victoria to be welcomed by her British garrison. This is the right Irish programme. We are sure it will be the programme of the United Irish League and of all Nationalists (*The Irish People*, 7 April 1900).

In a similar vein, it was pointed out in *The Freeman's Journal* that, 'there were... few cheers in the streets... the pavement, save where it was occupied by Belfast excursionists was singularly cool and undemonstrative... there was no enthusiasm. Hats and caps remained firmly fixed upon unsympathetic heads', (*The Freeman's Journal*, 5 April 1900). More militant nationalists went on to channel their opposition and resistance to the royal visit by organizing a counter-demonstration in the form of a torchlight procession on the night of the royal party's arrival in Ireland, as well as a party for children that would serve as a counter to the two 'children's treats' hosted by the Queen during her visit. The 'Irish patriotic children's treat' was organised by the leading republican, Maud Gonne, and Inghinidhe na hEireann (Daughters of Ireland) and was staged some weeks later on 1 July. The event was 'from its inception, a decidedly political and counter-hegemonic gesture' (Condon, 2000, p. 173) and it proved to be the most celebrated event staged in opposition to the Queen's visit. In many ways these counter demonstrations and events provided nationalists with a means of demonstrating their resistance to the royal visit and all that it represented. The organization of the torchlight procession, of a march commemorating the centenary of the Act of Union and of the patriotic children's treat went some way towards destabilizing the official narrative of loyal support and warm welcome. While countless Dubliners did indeed line the lavishly decorated streets to the city to cheer and welcome the royal party, contributing to the outward exhibitionism of Irish loyalty to crown and empire, the rituals of resistance that marked this event, served to underline the contested and hybrid nature of Dublin at the turn of the century. Although they largely failed to ignite a great deal of popular support, and while the impact of provocative journalism and *ad hoc* demonstrations was quite limited, such efforts did establish a network of anti-imperialist contacts that were to be further consolidated during following royal visits. When King Edward arrived in 1903, for example, he encountered a vocal, well-organized and decisive opposition and it is striking that although Dublin was once again lavishly decorated for his visit he was not presented with a formal address of welcome by Dublin Corporation. In many ways the lack of success of opposition groups exposed the limitations of conventional Irish nationalism and forced the creation of a more consolidated nationalist opposition. During that visit King Edward travelled the length and breadth of the country and an imperial veneer was once again painted upon urban and rural landscapes alike. The warm reception he received in Belfast, where he was presented with a formal address of welcome and unveiled a statue of his late mother, Queen Victoria, set the tone for the rest of his time in Ulster and stood in marked contrast to the scenes of well-organized and fervent nationalist and republican resistance that prevailed in the rest of the country.

Conclusion

The royal visits to Ireland at the turn of the century proved significant for a variety of reasons that relate closely to the politics of power that prevailed at this 'contact zone'. During each of the five visits the cultural landscape was effectively transformed and the royal processions through the city of Dublin as well as elsewhere were closely connected to the efforts of the Dublin Castle administration to cultivate a sense of imperial identity and draw citizens closer to empire. The pomp and ceremony, the elements of display and theatricality, the emphasis on the visual and the aural, all combined to create a spectacle of awe and wonder, a 'web of signification' that was cleverly spun by the state and supported by loyal institutions. Each visit made use of the very public dimensions of display and theatricality in an attempt to legitimate state power and cultivate a narrative of imperial identity. On each occasion the authorities succeeded in capturing the attention of the citizens. The ceremonial parade through the city, the symbolic events of welcoming, the official address, decorations, temporary structures, lights and fireworks, as well as the unique soundscape of a royal visit and the military presence, all played their part in this ritual and helped create the imperial veneer that transformed the city. Beneath the bunting and the Venetian arches these visits attempted to cultivate a sense of loyalty to empire, to draw citizens into the imperial project and perpetuate its ideology.

However, unlike other truly imperial cities where such displays of choreographed theatre proceeded smoothly and without debate, and where loyalty was more easily cultivated, Dublin and Ireland were caught in something of a schizophrenic position which these visits actually brought into sharp focus. This brings us back to a question earlier raised: did these visits confirm the city's status as imperial partner, the second city of the Empire or did they underline the fact that Ireland was a colonial subordinate? Queen Victoria's visit met with a significant degree of opposition from subordinate groups of nationalists who struggled to establish their own sense of identity. This opposition escalated with each succeeding visit in tandem with broader political developments that saw Ireland inch closer to Home Rule. So, while each visit succeeded in creating the veneer of an imperial city, they were less than successful in cultivating among citizens a more long-term sense of loyalty and imperial identity. These spectacular events effectively galvanized nationalist groups into opposition activity, as they set about rejecting all that the spectacle of the royal visit represented. In fact, the appearance of the British monarch in the city brought issues of national identity, self-reliance and political independence into sharp focus, just as much as the issue of loyalty to empire.

References

Bartlett, T. (1988), 'What Ish My Nation?: Themes in Irish History, 1550-1850', in T. Bartlett *et al.* (eds), *Irish Studies: A General Introduction*, Gill and Macmillan, Dublin, pp. 44-59.

Bell, J. (1999), 'Redefining National Identity in Uzbekistan', *Ecumene*, vol. 6, pp. 183-207.

Condon, J. (2000), 'The Patriotic Children's Treat: Irish Nationalism and Children's Culture at the Twilight of Empire', *Irish Studies Review*, vol. 8, pp. 167-78.

Davis, S.G. (1986), *Parades and Power. Street Theatre in Nineteenth Century Philadelphia*, Temple University Press, Philadelphia.

Gibbons, L. (1998), 'Ireland and Colonisation Theory', *Interventions*, vol. 1, p.27.

Goheen, P. (1990), 'Symbols in the Streets: Parades in Victorian Urban Canada', *Urban History Review*, vol. 18, pp. 237-43.

Goheen, P. (1993), 'Parading: a Lively Tradition in Early Victorian Toronto', in A.R.H. Baker and B. Gideon (eds), *Ideology and Landscape in Historical Perspective*, Cambridge University Press, Cambridge, pp. 330-351.

Goheen, P. (1994), 'Negotiating Access to Public Space in mid-19th Century Toronto', *Journal of Historical Geography*, vol. 20, pp. 430-449.

Goheen, P. (1998), 'Public Space and the Geography of the Modern City', *Progress in Human Geography*, vol. 22, pp. 479-96.

Fitzpatrick, D. (1998), *The Two Irelands, 1912-1939*, Oxford University Press, Oxford.

Fraser, T. (2000), *The Irish Parading Tradition: Following the Drum*, Macmillan, London.

Howe, S. (2000), *Ireland and Empire: Colonial Legacies in Irish History and Culture*, Oxford University Press, Oxford.

Jarman, N. and Bryan, D. (1998), *From Rights to Riots: Nationalist Parades in the North of Ireland*, University of Ulster, Coleraine.

Kennedy, L. (1996), *Colonialism, Religion and Nationalism in Ireland*, Institute for Irish Studies, Belfast.

Kenny, K. (eds), (2004), *Ireland and the British Empire*, Oxford University Press, Oxford.

Kong, L. (1993), 'Political Symbolism of Religious Building in Singapore', *Environment and Planning D: Society and Space*, vol. 11, pp. 23-45.

Marston, S. (1989), 'Public Rituals and Community Power: St Patrick's Day Parades in Lowell, Massachusetts, 1871-74', *Political Geography Quarterly*, vol. 8, pp. 255-69.

Paseta, S. (1999), 'Nationalist Responses to Two Royal Visits, 1900 and 1903', *Irish Historical Studies*, vol. 31, pp. 488-504.

Yeoh, B. (2000), 'Historical Geographies of the Colonised World', in B. Graham and C. Nash (eds), *Modern Historical Geographies*, Pearson, Harlow, pp. 146-66.

Yeoh, B. and Kong, L. (1997), 'The Construction of National Identity Through the Production of Ritual and Spectacle', *Political Geography*, vol. 16, pp. 213-39.

PART II
'TRANSLOCATIONS'

Chapter 7

Environment-identity Convergences in Australia, 1880-1950

J.M. Powell

It has been widely assumed that early non-indigenous perceptions of Australia were warped by the application of familiar British lenses to 'antipodean perversities' characterizing the fauna, flora and climate of the new country. According to some persisting analyses, the loss of the 'human associations of an historic past' aggravated the disorientation (Smith, 1975, 129; *cf.* Martin, 1993). Gradually, the 'whining song of the exile' was reportedly countered by balladists, poets and writers, by a distinctively Australian school of landscape painting in the 1880s, and in the generation of nationalist sentiment accompanying the move towards federation, achieved in 1901 (Ward, 1958; Serle, 1973; *cf.* Powell, 1977, pp. 21-82). These treatments concluded that the new aesthetic substituted sun-filled vistas of sheep-and-wheat country for the assertive deserts, the 'funereal' forests, and anxieties over the conjectured continuum of 'primeval' Nature and the supposedly savage state of culture it had fostered – exposing the latter as disagreeably reminiscent of a posited 'degeneration of species' in New World settings (*cf.* Powell, 1977, pp. 29-30). Accompanying the change and fostered by it, projections of 'progress' sketched 'Man above Nature', declared the conquest of unhappy and stressful pasts, and pointed to a confident future: economics in drag.

As the first section of this paper begins to attest, the inescapable claims of science and technology make that familiar narrative rather less convincing. In addition, its preoccupation with ingrained British sensitivities underestimates other psychological and social origins of a common (but scarcely universal) reluctance to 'embrace the bush' (Bonyhady, 2000). The second section notes that this was the kind of displacement in which aggressive hostility co-existed with abject fear, and patently, where perplexity linked and intensified those emotions, accelerated understandings could raise the comfort level quite dramatically. The older explanations seem to have oversimplified that context. Recent writings strive to demonstrate that too little is known about the very construction of the stubborn negatives, about the ways in which it was possible to dismiss or tolerate them, and how they could be turned to better account. It is being shown once again that environmental enigmas, constraints and opportunities have been central to the quest for a secure coming-to-terms in Australia: earlier historical-geographical research had concluded that environmental learning, based upon widespread empirical testing, science, folk geography and the landscape authorship of key resource managers, had crucially assisted the increasingly place-engaged art and literature to contribute towards a process of 'settling' which seemed authentically Australian (Meinig, 1962; Heathcote, 1965; Powell, 1976, 1977; see also Williams, 1974). So the principal objective of the following discussion is to consult a selection of complementary modern assessments of the interactions between resource appraisal, changing attitudes towards nature, and the problematical 'sense of belonging'. The brief survey suggests that the recognition, utilization and protection of inter-related, emphatically distinctive endowments created a validating interest and stimulus which also served to articulate a more confident differentiation from metropolitan Britain.

Watering

Towards the end of the 19ᵗʰ century, after a mere hundred years of white settlement, Australia's human residents were quite tightly concentrated into its temperate and 'Mediterranean' fringe. The vast interior was still lightly used and lightly occupied. The fringe accommodated the most productive land-uses and its constituent landscapes – the comparatively intensive sheep and beef zones, the wheat-and-sheep belts, dairying enclaves, and so on – recorded a collective engagement with nature in which the exploitation of water availabilities had been a central task. Nothing less might have been expected for the driest of the inhabited, vegetated continents, yet the story of that pivotal engagement has been generally taken for granted. The Australian experience abounds in water marks: they afford a peculiarly reliable warranty of authenticity. Their insertion at this introductory point balances the following sections with a reminder that modes of environmental learning, like environmental mistakes, were inextricably tied to a powerful development imperative.

Nor should they be divorced from a consuming quest for an improved social contract in the new country, arguably most tangible and problematical in the pursuit of idealistic programmes for sequenced, equitable disposals of the public domain. Those programmes derived from a style of colonial socialism which included acceptance and indeed expectation of government intervention; typically, it gave rise to changing conceptualizations of graduated, pre-settlement property sizes reflecting social equity and environmental capacity notions, with related adjustments in complex tenure arrangements. That underlines the critical intermeshing of resource appraisal, conservation and equity considerations in contemporary development planning. Its pervasive influence was strengthened by the evolution of a kaleidoscope of 'welfare state' measures, but for our present purposes the single example of water management must suffice. Water management was implicated in many defining moments which simultaneously displayed local ingenuity and opened Australia to the wider world. For example, during the gold rushes of the 1850s, recognition of the singular importance of water supplies eventually led to the design of novel legislation to block ambitious individuals from controlling the key resource. In this regard and in the delivery of water by means of makeshift races, flumes and aqueducts, Californian miners showed the value of prior lessons and the relevance of the British inheritance was sharply interrogated. Over the same period, booming townships faced recurrent crises over cholera and typhoid. Melbourne's water managers responded by pioneering a robust concept of 'closed catchments' – dedicating large, sacrosanct, tree-covered reserves to ensure the harvesting of high quality supplies. They remain largely intact today. And to the west and north of Melbourne, goldfields municipalities combined in a muddled but surprisingly innovative small basin project for multiple-purpose development. In the last quarter of the century, Victoria's Minister for Water Resources, Alfred Deakin, returned from an overseas mission with plans for irrigation developments in the dry north and northwest of the Colony. The Victorian programme drew diligently upon Indian, European and Western American experience, including John Wesley Powell's 'watershed democracy' notion, in introducing a series of decentralized Trusts to supervise the effort. Inspections of the failure of the Trust system registered the need for preparatory environmental assessment. It also led directly to the establishment in 1905 of a centralized State Rivers and Water Supply Commission with a monopolistic purview over non-metropolitan supplies, and for the supervision of a good deal of new settlement expansion. Its bailiwick covered an area almost as large as the United Kingdom and it would be parochially lauded as an anticipation of the Tennessee Valley Corporation: evaluations of the American connection commonly expressed a regard for the practical relevance of prior 'frontier' experience, affirmed during the emergence in the United States of a trumpeted 'nurturing state' in the New Deal years (Powell, 2000). In 1907 affirmation took the form of the appointment of Elwood Mead to head the new Commission. Mead's executive background in the mundanely-named Irrigation and Drainage Investigations Bureau of the US Department of Agriculture was highly pertinent; just as significantly, however, after his return to the United States in 1915 he experimented with a little Victorian-style 'socialistic' planning in California – one of the lesser known reciprocals in our absorbing history of Pacific transactions (Powell, 1982, 1989; *cf.* Tyrrell, 1999).

From the late 19th century until the end of the interwar period, the discovery and exploitation of the Great Artesian Basin secured the development of an enormous region as extensive as the combined national territories of France and Spain. In characteristically Australian fashion, the related technological and organizational infrastructure was an enterprising mixture of contemporary American, European and local approaches. Populist speculation on the origins of this mysterious variant of the old 'inland sea' gave rise to precocious displays of vernacular environmental thought before the riddle was solved. Further south, perceptions of an intimate connection between regional development, national security and changing balances in world power were also advancing water management in the public imagination. During and after World War II the massive, multiple-purpose (especially power and irrigation) Snowy Mountains Scheme brought a clamour for foreign investment that justified federal leadership. So did defence considerations, revived interstate jealousies, legal tangles and a new push for non-British immigration that proved quite portentous. Construction commenced in 1949 and was completed in 1974.

Casual encounters with the evolving story can give the impression that the construction effected a revolutionary change in the appreciation of a hitherto little-known 'High Country'. In fact a small but determined alliance of competent ecologists, foresters, soil surveyors, bushwalkers and government land managers had successfully pressed the New South Wales Authorities to proclaim a Koskiusko State Park in 1944, mainly to curb the grazing and burning operations on summer leaseholds. In some technical, organizational and psychological aspects, the giant scheme fitted into an international group of wartime and postwar extravaganzas which included the Grand Coulee, Damodar and Kariba Dams, and although its detailed methodology called upon hard Australian lessons it was also strongly influenced by the precedents set by the New Deal's Tennessee Valley Authority. The impressive resonances galvanized the small, isolated, vulnerable nation and the scale of the undertaking could not fail to foster a new level of environmental curiosity attention. It sowed the seeds of a vigorous critique. Construction work pitted the Authority's leading personnel against the romanticized traditions of the upland cattlemen. Dam building, reservoir protection and related activities stepped-up the investigation of hydrological systems and soil mechanics within the region, and opened it up for an army of naturalists, tourists, walkers, climbers and skiers. Controversy intensified over the impact of grazing on 'alpine' vegetation – its premium inflated, naturally enough, by attributions of exceptionality in a continental preponderance of limited relief. Towards the end of the 20th century the pride the Scheme had understandably inspired would be judged misplaced by economists and environmentalists.

But in fact the water mark had already become prominent at the level of national government from the very inception of federation in 1901. Although the states had retained responsibility for natural resources, federal authorities were soon obliged to intervene in drawn-out wrangling over the management of the River Murray for navigation, water supply and irrigation. The initial outcome was a good illustration of what would become a familiar accretion of federal powers – part stealth, part incidental. In 1915 a conciliatory *River Murray Waters Agreement* set up a River Murray Commission staffed and financed by the South Australian, Victorian, New South Wales and Commonwealth Governments. It lasted until the 1980s, by which time it was realized that the Agreement had produced an inefficient southerly bias in the management of a much larger and increasingly troubled region – the Murray-Darling Basin, commanding a series of catchments which stretched into Queensland. The hopeful solution to inherited 19th century problems was now considered implicated in the failure to recognize the worth of holistic or integrated management for inter-related tasks associated with salinization, soil erosion, land use planning, water pricing, drainage, improved community dialogue and other Basin-wide problems. But crucially, in the present context, the 70-year prelude to that realization had been punctuated by soul-searching on the newsworthy theme of environmental responsibility from a host of farmers, graziers, planners, scientists and others. It built the necessary platform for a comprehensive overhaul based on 'bioregional' initiatives which gave Australia an elevated status in water management circles, and its history cultivates the patriotic conscience. Australians were urged to see that their 'MDB' – the heartland of much of the nation's art, literature and rural production – had been made hostage to the environmental mistakes of the past. Its survival depended upon improved understandings of that legacy, interdisciplinary engagement, and resolute community mobilization (Powell, 1993a).

Exorcism

For much of the colonial period, the Australian landscape must have represented the very opposite of the fortuitously available 'vast, empty spaces'. The slightest acknowledgement of its antipodean characteristics could bring evocations of inter-racial hostilities and the shameful legacy of convictism. Following Smith (1975) and Hirst (1983), Griffiths (1996) and others have explored some of the ways in which the effects of psychological trauma severely tested the place-making, past-using functions we attribute to the human condition. It is not necessary to allude to 'coping strategies' and the like to note a few of the responses relevant to the present review. Australia was haunted by Aborigines and convicts. Conspiracies of silence were encouraged by borrowings from Social Darwinism promoting the inevitability of the disappearance of 'native' populations and the advance of 'superior' white civilizations. Griffiths' *Hunters and Collectors* (1996, p. 108) writes of an 'unconscious reflex, an unobserved accretion of silence', while Reynolds (1989, 1990, 1992) has shown how the colonists drew a curtain over the past, veiling the emotions. While memories festered and befuddled, the inscriptions of government cartographers provided conscience-jabbing titles.

Griffiths (1996) reports that the guilty silences were transmogrified into kitsch sentimentalities during the last quarter of the 19th century, when wretched remnant individuals were given some patronizing recognition – 'King Billy', 'Queen Aggie' and so on – and local worthies campaigned for the erection of modest memorials to mark the deaths of the 'last of the clans' in their respective districts. Abundant work there, surely, for honest contemporary scholarship, yet how would that have been accomplished without giving offence, specifying culpability? Frontier violence had not entirely escaped the early chroniclers, but occasional exposés of white savagery were publicly reviled as the slanderous charges of race traitors. During the later period of memorialization one noted Victorian ethnographer, James Dawson, raged against the murderous 'extirpation', but his fellow pioneers refused his requests for contributions towards the erection of a suitable obelisk precisely because it would stand as a permanent accusation.

Paranoia about the convict stain distorted early scholarship in New South Wales and Tasmania, which had received the majority of the 'transported' immigrants. Material evidence of convictism, including written documents and hated buildings, was frequently destroyed. Darwinian notions might be co-opted to soothe reflections on the 'passing' of the Aborigines, but the same muddled reasoning could raise doubts about the 'calibre' of settler stock. One way of dealing with this involved a twist on the 'dying race' theory: commentators suggested that the convicts, like the Aborigines, would be supplanted by superior types as civilization progressed (McGregor, 1997; Anderson, 2002; for a wider imperial setting *cf.* Trainor, 1994). Another approach was to appeal for an end to the transportation system so that Australian history could be allowed to begin again on a clean slate. That incorporated a denial (or 'suppression') of the less inhibited sense of Australian identity developing amongst children born in the adopted country. Inglis (1974, 1985) and others take this idea of a general need to defuse the past much further. Unfortunately, in this view, what was passionately desired was a massive and definitive proof of genetic quality. Hence, perhaps, the extraordinary interest in sporting contests, at the imperial level often intriguingly referred to as *tests*, and especially the excited rush to war when the chance was presented. Australia's 1885 Sudan contingent was warmly applauded by the children and grand-children of convicts. The Great War itself was widely accepted as a dual test – of race, of character. Even today the contemporary logic seems at once irresistible and painful: with the violent past expunged, future-orientated Australians had plumped for a *history-making* conflict. And so to the horrors of Gallipoli and the Western Front.

This tale of progress towards a kind of exorcism is but one thread in a challenging tangle. Never unequivocally freed of their Old World origins, settler societies had to find ways to put down roots, and that was a central aim of this history-making and place-making. It is very much the business of their inheritors to take heed of these efforts, whether splendid, clumsy, deeply embarrassing or repugnant, and to reflect on our own husbandry. But too much of this section depends on the getting and blurring of memories. Other strands in the skein came from advocates of alternative reckonings on the unique qualities of the Australian experience. Natural history writing, memorialization activities and the beginnings of a form of conservationism tweaked the focus on the qualities of place.

Selbornianism

Convenient 'slicing' of 19th century Australian history tends to accentuate apparent breaks in the 1850s and 1880s, but it is not entirely at odds with contemporary reflections on the settlement experience. Although there is hard evidence of important continuities and regional variations, incisive transformations certainly followed the introduction of gold-mining in the 1850s and could not fail to stir the memories of those who had been involved in the making of the earlier 'squatting' or pastoral frontiers. By the 1880s, the immigrant pioneers of the golden fifties were contemplating another displacement, by their own Australian-born children. As Davison (1989) notes, in each group the consequent arousal of past-consciousness offered fertile ground for local history-writers as well as for the picking-over of shared reminiscences at the familial and neighbourhood level. The same process encouraged the formation of historical societies, old colonists' associations and the like, and in some degree the effect was to promote or confirm a striking of community roots – home-grown, for home-growing.

In the literary field, a succession of popular writers quarried the accessible past and ventured an interesting chronology. Much of this writing was conventional and melodramatic, but it helped towards a new evaluation. The main characters and settings became frontier stereotypes: the explorer, pioneer pastoralist, the goldfields (and therefore the miner or 'digger'), even the escaped convict. Bushranger histories offered spice but also a worrying odour of convictism. Exploration histories might say something about individual ennoblement, but they encouraged a much more inclusive advance on the wilderness. A little balance was added by the emergence of another *genre*, the development (or progress) narrative relating the spread of settlement and civilization. An important breakthrough came with elevations of the contribution made to the formation of authentic national character by the pioneer farmer and his literary companion the protean 'bushman' – the latter signifying a drover, shearer, bullock-driver, timber-worker or similarly romantic rural labourer of any type. This promotion is most closely associated with the spate of short stories, ballads, poems and rough-house opinion churned out by the radical-nationalist *Bulletin* magazine. Commencing publication in 1880, the *Bulletin* had built a circulation of 80,000 by 1890; it fostered the careers of several of Australia's most popular writers, including Henry Lawson and 'Banjo' Paterson. Australian contributions to the 'natural history' genre were similarly, in part, the products of nostalgia; in addition, however, they drew inspiration from the emergent natural sciences and from the growing notoriety of Darwinian themes. The early advertising of colonial Australia as a 'hell on earth' had been deployed in the cause of social control in Britain, but in scientific circles such fabrications could never compete with images of an antipodean treasure trove. Actual and hinted connections to Darwin, the Hookers of Kew Gardens, Joseph Banks and other British and European scientists put high premiums on original research in Australia (*cf.* Martin, 1993; Gascoigne, 1994). The lively local reception of George Perkins Marsh's *Man and Nature* reflected a natural history enthusiasm which had been excited by the circulation of *Origin of Species*, and the related critiques of environmental appraisal from prominent earth scientists-clerics proved quite insightful (Powell, 1975, 1976). But recent revisionism also reports rising participation in amateur natural history during the late 19th and early 20th centuries, and it is no longer safe to dismiss this as 'antiquarian' or 'peripheral'. The tendency is rather to acknowledge the presence of alternative or at any rate parallel streams of consciousness alongside the scientific and scholarly lines. There are some excellent examples. Gilbert White's *Natural History of Selborne* had first appeared, coincidentally, in the first year or so of European settlement in Australia, and proof of its continuing impact can be traced throughout the period under review. Thoreau's *Walden* was also robustly influential. In their different ways, each of these great classics supplied a model for urban intellectuals who felt disturbed by the 'passing' of the wilderness and the reassuring rhythms of rural life, and were ready to detail their own place-attachments. That rising sentiment seemed more than a match for broad-brush negativity.

Tom Griffiths (1996) revisits some impressive natural history writers in Victoria. There was, for example, Donald Macdonald and Charles Barrett, who reached comparatively wide audiences with their short essays and newspaper columns between the late 19th century and the interwar years. Macdonald occasionally crafted pleasing rural idealizations into a distinctive style of reporting on cricket matches – an anticipation, perhaps, of Britain's John Arlott after World

War II – and for its time, this unusual avocation won an appreciative following. His affectionate recollections of village life were spiked with alarms about the threats posed by technological change and the expansion of Melbourne (Macdonald, 1887), and in the latter stages of his career he converted his newspaper columns into regular exchanges of queries and field observations from his keen correspondents. Macdonald beckoned Charles Barrett, another journalist, along the same path, and the protégé ventured much deeper, exploring the connections between natural history and the cultivation of national identity.

Trained to more exacting tastes, later Australian scholars bypassed the efforts of those who tried to cultivate a sense of curiosity and concern about land and life in the adopted country. In this context too much has been made of the initiatives of Melbourne's 'Heidelberg' school of painters in the 1880s and 1890s. From their charming camps on the edge of the city, the Heidelbergers seemed to throw off the inhibitions of previous generations, experimenting with recognizably Australian colours, shapes and light. If, however, it made them teachers in an accelerated learning process, then that impact was not widely registered until, say, the second decade of the 20th century. Griffiths believes that Macdonald, Barrett and other popular writers had prepared the ground. Interestingly, some of the latter set up their own bush camps outside Melbourne: one of them was actually christened 'Walden'. A regular associate, Claude Kinane, was inspired by the work of Britain's pioneering nature photographers, C. and B. Kearton, to do the same for Australia. Whether subtly insinuated or (especially in later years) self-consciously proclaimed, the promotion of nationalism is an inescapable characteristic of these works. At one level, they campaigned for the invention and adoption of vernacular (they said 'popular') names for fauna and flora to enhance the mission: patriotism either required or was to be expressed in a love of the bush. Macdonald's experience as a war correspondent in South Africa led to the publication of nationalistic accounts, and to those clubby nature columns noted above, which were also conduits for contemporary anxieties over urbanization and industrialization. Both Barrett and Macdonald wrote for Australian boys and lectured on 'bush-craft', and like most of their group they quietly aligned themselves with the nationalistic contributors to the aggressive *Bulletin*.

If, as Griffiths (1996, p. 150) so nicely puts it, the natural history writers, artists, balladists and novelists invited an 'imaginative appropriation' of the new country, Australia's 'progressives' pressed instead for a steady inculcation of citizenship in and beyond the schools, employing very practical recruitments of the past to serve as moral inspiration (Roe 1984; *cf.* Freestone, 1989). Links with Selbornianism are best seen in the identification and commemoration of significant sites and places. As in the United States, cultural anxieties led to *monumentalism*, a form of topographical appreciation that evoked nationalistic sentiment. Where the features seemed grand and agreeably Australian they might be set aside ('declared') as discrete monuments: bizarre rock formations, the tallest and oldest trees in any forested district. Other places nobly 'associated' with the achievements of the white explorers and the earliest of the pioneer settlers were also monumentalized as cues for place-bonding exercises. Cairns and obelisks marked the routes taken by the explorers; plaques were affixed to culturally significant buildings. Lovingly planted avenues of honour remembered local *sacrifices* of soldiers in the Great War: simple statues of combat men astride each appalling, tolling list of victims continue to invite sober consideration, all too often in the merest hamlets or villages. So the country's landscapes (including its townscapes and roadscapes) were made more didactic: stories were spatially harboured, popular attachments nourished (Inglis, 1987; *cf.* Jeans, 1988; Jeans and Spearritt, 1980). Proof of possession seemed to demand such instruments.

As in all settler societies, increased spatial and social mobility became relatively common throughout Australia, and the attainment of truly intimate Selbornian relationships with place might well have been regarded as another back-burner option. At minimum, Griffiths' *Hunters and Collectors* establishes that the influence of the natural history writers merits more probing analyses. Other research re-connects them with prominent individual scientists. For example, just as a number of natural historians earnestly hoped to become 'more Aboriginal' – that is, more in tune with Australia – so one extraordinary all-rounder in science, Alfred Howitt, whose practical and scholarly achievements were nothing if not expressions of his own yearning for *adaptation*, briskly described himself as 'naturally a savage' (Powell, 1993a, p. 52). But thus far, the settled fringe has dominated our survey and the local and the regional have been privileged. In the early 20th century, the continent's mysterious centre became a special, sacred space for natural history authors and other Australians. The suitably-labelled national magazine *Walkabout* offered an intensified celebration of Outbackery and encouraged pilgrimages to the inland. It was surely one of the better antipodean quirks to locate a rousing frontier at the hub rather than at the edge of the national space: the 'desert-as-howling-wilderness' or 'desert as desolation' theme was said to belong with the rest of the European cast-offs, and it began to lose credibility as the new nation improved its grip on a purloined continent (Haynes, 1998). Increasingly, the 'Dead Heart' became conspicuously depicted or even routinely apprehended as whitefella sacred space, the 'Red Centre'. It throbbed with life for a new breed of field scientists. For others, journeying to the very Heart of the country was often portentously conceived as a psychological expedition, a rite of passage.

Faunal

Although there were interesting exceptions to the rule, on the whole the early invaders – definably 'feral' in lights denied them – were inclined to position Australia's unusual wildlife at the furthest extreme of *exotic* before that term accrued positive associations. Imported benchmarks of interest and beauty did little or nothing to subdue alarming encounters with bizarre features and behaviours. The eerie nocturnalities of the bush deepened the profundity of the silences noted by Griffiths, increasing the distance between human and nonhuman species. As opening confusions gave way to degrees of indifference and hostility, indigenous fauna came to be coarsely undervalued as 'useless', 'weird', and 'vermin', some barely qualifying as 'game'. In such situations even the most earnest tracking of ramshackle and disregarded animal protection legislation brings meagre returns, but that is less true of developments in the review period. During the closing decades of the 19th century the familiar tension between overseas and local concerns enlivened a more determined but still convoluted discourse over the fate of introduced and native creatures. Reflecting, as ever, the different experiences and natural endowments of the individual political jurisdictions, by the late 1940s the protection statutes had achieved something approaching a common set of approaches which accommodated recognizably modern gestures towards the ethical resolution of competing utilitarian, ecological and aesthetic arguments. In its essentials, this uneven learning process was ineluctably Australian. It would remain so.

When Science's specialists pursued other agendas, the liveliest of Australia's amateurs worked with a will to direct public attention to the fate of imperilled fauna. Some of their activity was designedly parochial – keen ornithologists, for example, were influential in ensuring the teaching of Nature Study, in founding the Gould League of Bird Lovers (named for a prominent pioneer naturalist) and in the promotion of Bird Day to accompany the existing 'Arbor Day'. By the mid-1930s the Victorian branch of the Gould League alone boasted some 100,000 members (Robin, 2001). The product was a legion of potential volunteers for local operations. But it is seldom fruitful to draw hard and fast lines between parochial, national and international trends. Fears of global 'extinctions' of any kind, for instance, whether or not they were allowed scientific accreditation, could often be relied upon to incite Australian investigations on home territory. Until recently, developments in forest, soil and water conservation have attracted more interest in Australia's under-subscribed history of resource appraisal and environmental management. Those tasks are far from completed, but in comparison with investigations of the fate of the Australian fauna they are well advanced. It is time to redress the balance and indeed to co-ordinate the various lines.

Stubbs' (2001) treatment of broadly representative trends in New South Wales finds this overlapping sequence: assorted protections for imported game in the 1860s; favouring of aesthetically significant native birdlife in the 1880s and 1890s; the inclusion, especially from the early 20th century, of certain commercially useful marsupials and those species threatened with extinction; and finally a long transition to the introduction of combined legislation, in the late 1940s, for the protection of *all* birds and animals with the exception of specified exclusions. All through the messy protraction, the laws were rather difficult to make and exceedingly difficult to interpret and enforce. In particular, they collided with the vested interests of cultivators and graziers, and laboured under imputed attachments to real and supposed differentiations between urban and rural sentiments: reminiscent of a jargonized lifestyle choice, 'Sydney or the Bush', they succinctly acknowledged a two-Australias divide.

In support of that central contention, it has been found that, during the first third of the 20th century, a few species appeared to be selected for veneration as convenient denotations because of their 'claimable' characteristics, and that these changes can be traced to a complex blending of influential social movements anchored in the larger urban centres (MacCulloch, 1994). 'Australian' came to be symbolically represented by the kangaroo, emu, koala and other indisputably antipodean creatures. Two of the leading movements, initially quite discreet entities, pressed for animal *protection* and the *preservation* of endangered species. They helped to produce innovative (if only partially effective) legislation, in conjunction with modifications in school curricula, to lay the basis for wider disseminations. Animal protectionism drew on evangelical Christian views and was inseparable, socially and intellectually, from a broader 'reformist' framework encompassing strains of sabbatarianism and temperance for 'self-improvement'. At the outset, it was more interested in the consequences of the human-dependency of domestic pets and farm animals. Cruelty was inhumane, compassion its opposite. A similarly anthropocentric preservationist impulse fed upon scientific and historical associations which voiced well-informed fears about the destruction of the natural environment and threats to cherished features in the built environment. Together, these protesting minorities succeeded in exposing and arguably exacerbating the urban-rural split.

In addition these movements picked out some of the ways in which urban Australians had become even more fully recruited by, or fancied themselves more in step with, the wider world. Both were influenced by earlier and parallel developments in Britain and, to a lesser extent, in the United States, but more proximate troubles had raised their hackles. Feral rabbits had usurped millions of small marsupials, drastically reduced the productivity of countless acres, triggered soil erosion, destroyed livelihoods. A huge population of feral pigs rampaged through paddocks and bushland – by the end of the 20th century it would exceed that of the human population – the countryside suffered recurrent plagues of mice, locusts and caterpillars, and blackberries, thistles, sweet briar and other 'noxious weeds' invaded the humble backblocks. Fingers burnt, financially pressed, the well-meaning Acclimatization Societies retreated from releasing further exotic species, concentrating their energies on more manageable assets in their zoological gardens. Like their American counterparts, they switched to the cheaper expedient of collecting local animals in connection with a policy of interchange with overseas institutions, but without pleasing many protectionists or preservationists. International influences were more keenly felt after the founding of wildlife protection associations in Europe around the beginning of the 20th century, and following the introduction of a series of ambitious convention dealing with 'game' animals in Africa. In Australia, conservationists and others also became interested in the application of the principles of the Society for the Preservation of Fauna of the Empire, started in the United Kingdom in 1903 (de Courcy, 1995; Mazur, 2001).

From the 1890s until about 1920 a boisterous 'anti-plumage' campaign helped towards some blending of the two main camps. The campaign focused on the international issue of 'murderous millinery', especially the use of feathers and even stuffed birds (whether whole or part!) to adorn women's hats and other clothing. Australia's spectacular birdlife was being plundered. Taking up and personally contributing to recent British and American trends (Robin, 2001, pp. 79-109), local protesters emphasized the barbarism associated with plumage collection, using graphic photographs and emotional descriptions to appeal directly to the higher maternal instincts of women, and astutely elaborating on the economic utility of insectivorous species and the enviably unique characteristics of Australian birds. That was much easier than attempting to corner the predominantly male hunters and farmers adjudged responsible for the despicable trade. The strategy directed the public to those species most at risk – the superb lyrebird, herons and egrets, a number of the multi-coloured parrots, for instance – and effectively used metropolitan and country newspapers to politicize the murderous millinery issue. It also urged the federal government to intervene by using its constitutional powers over the regulation of imports and exports. Discomfited scientists eschewed the strident propaganda, but frequently supplied critical information to the preservationist cause. Too many of them withdrew into the laboratories and lecture theatres, preferring to work through their professional organizations to provide conservation-related methods and findings to the governments of the day. So the public looked instead to 'amateur' associations. At the risk of underestimating the contemporary continuum of amateur-professional allegiances, it does seem that the ramifications of this unfortunate estrangement blighted the remainder of the survey period, and that Australian environment was the loser.

In the case of mammals, the rural lobby maintained its outrage against the dingo's predation on livestock, and against all apparent competitors for grazing resources, notably plains kangaroos and the emu. While protectionism *per se* was usually fully committed in the populated fringe, the preservationists entered the fray with an attempt to stress the need for 'sustainability'. This was backed by scientific concerns that ill-informed reactions to short-term population explosions had not been factoring in the effects of natural recovery mechanisms following devastating droughts, fires and floods. In any event effective regulation was impossible, given the sheer size of the country and the concomitant problems of policing its remote, sparsely populated districts. Perhaps the greatest resistance came, however, from the familiar compound of organizational inheritance, psychology and economics. Outback graziers, supremos on most of the decentralized Pastures Protection Boards, trumpeted their inalienable right to safeguard the territory wrested from harsh frontiers and brushed aside the indignation of softer urban brethren. In turn, their parliamentary representatives argued that native animals competed for scarce water and fodder, and presented petitions for the lifting of close seasons which had been introduced to maintain nominated species. In the absence of thorough scientific tests at each local level the graziers' claims withstood serious challenge. Where their resistance appeared a little shaky or hinted at self-serving, a subsidiary argument was often raised: opportunities in furs, kangaroo-shooting, emu and dingo eradication and the like helped to keep needy Outback families above the margin.

MacCulloch's (1994) main argument is that preservationism succeeded best, in Australia as elsewhere, when it concentrated on *selected* species. Reduced populations of the platypus meant that it had become less directly threatened by commerce. The dingo, seemingly universally condemned, had been excluded from the pantheon. Political realities and rickety research findings meant that the 'opposum' (later simply 'possum') and kangaroo would have to make do with partial protection, and that would become notoriously so in the 'peripheral' states of Tasmania and Queensland, which stubbornly exercized the right to maintain any number of destructive habits. The koala or 'native bear' was raised more easily into a favoured position, largely because its diminished populations were persuading the fur trade to think again, and although this had brought renewed pressure on kangaroos and opossums, weary preservationists found solace in the 'saving' of the koala. Understandably. That stage was reached at the end of the 1920s. The previous decade had witnessed an unprecedented onslaught and the preservationists' campaigns had engaged the emotions of the nation.

Koalas had been tentatively protected in Victoria from 1896 and in New South Wales from 1903, but their large populations in the more recently settled areas of Queensland remained under threat throughout the early 20th century. The Queensland government suspended protection for six months in 1918 and again in 1927. During the first open season over one million koala skins were marketed. Presumably, many more animals were killed and pronounced worthless, and entire animal communities were disrupted: apart from all the indiscriminate trapping and poisoning, associated disregard for habitats prejudiced the recovery rate. By 1927 the approximate *commercial* kill reached 600,000 and even the narrowest protectionists had joined the protest in earnest. There was a good deal of lobbying of state and federal representatives, and leading lights in the Wildlife Preservation Society of Australia (established 1909) wrote vigorously for the magazines and newspapers of the day. Skilfully placed photographs offered what is now the immediately recognisable stereotype around the globe – the inoffensive sleepy 'teddy-bear', baby on board, ensconced in a gum tree. As MacCulloch (1994) continues, the tactic could not fail to erect a type of anthropomorphized hierarchy in the animal kingdom: the koala became a favoured *creature of culture*. Adroit extensions of this valorization declared Australian ownership and welcomed other wild creatures into a national iconography.

Only marginally involved, high or higher science is made to appear sidelined in that outcome. If such unfortunates as the dingo or the marsupial rat remained beyond the pale, it was of course a reflection of social acceptance, not of sophisticated ecological appraisal. But the narration understates the importance of applied science and technology, growing respect for the unflinchingly Australian characteristics of that trend in the public arena evinced in rising increments of state and federal funding, an intimate interdigitation of innovations stemming from the empirical testing favoured by farm families and the laboratory, fieldwork and extension services run by government agencies, and close interchange around the British Empire and with the United States, particularly during the New Deal years. Examples include pioneering in 'dry farming' techniques, investigations of critical trace element deficiencies in Australian soils, revolutionary applications of superphosphate, research into tropical diseases in human and animal populations, and internationally renowned biological control measures (Williams, 1974, pp. 292-332; Powell, 1991, pp. 121-194, 2003). The formal introduction of natural reserves for protection and preservation in the late 1940s gives a better citation of scientific participation in the recognition of habitat needs, and in part it was: by that time the informed idealism behind the creation and ongoing management of the great parks in Africa and North America was magnetic in professional circles.

But the Australian inscriptions had faithfully recorded an inherited jumble of local inputs, amateur and specialist. The very idea of reserving space in the public interest had been an integral part of settlement management since the convict era. Intermittently incorporated into arrangements for water conservation and land use classification, it was elaborated in the setting aside of Australia's own major parks from the late 19th century, and had featured in the introduction of simple fauna sanctuaries near state schools in the 1930s.

Settling

If nations are imagined communities, then the endeavours related here performed some useful conjuring. My brief outline merely sets out to locate some of the ways in which an immigrant nation is said to have progressed from uncertainty, fear and embarrassment about its adopted country to a point at which it was beginning to feel more 'at home'. Having found a measure of comfort in an uncommon patrimony, Australians were beginning to feel more secure about the business of place-engagement, or rootedness, but how deep had it gone? How transparent and how lasting, for instance, those memorializing attributions? Well might the government-funded gravestones of three mounted policemen, famously killed by a gang of bushrangers at Stringybark Creek, lament an 'ornament to his country' and 'a terror to all evil-doers'; nearby their more eloquent memorial, reportedly raised by wide public subscription and so prominently located as to be unavoidable, has been routinely described by generations of locals as 'Ned Kelly's monument'. Such is life, and history-making. So too of our changing relationships with the environment. Nature in Australia could at last give plenty of stimulus to all and sundry, and over the review period it seems to have been increasingly well accepted that many of its secrets were being directly and indirectly uncovered by the relentless search for productivity. Yet some misgivings remained about the proposition that a place that could at last be appreciated might not, in a sense, fully reciprocate – not, at any rate, with the kind of bounty capable of justifying all the emotional and intellectual investment.

The stories of enriched environmental and historical sensibilities caution us against reducing the complexities of the Australian experience to wallets and wickets. They sketch out an emerging compact of sorts, but how firm had it become? Their examinations might have been profitably extended to include impingements of the wider world on those complicated matters of 'belonging', into some of the more ambiguous effects of the addictive appeal of development, and especially into nation-wide purviews from notorious dissenters. What presents as a growing satisfaction derived from a range of local and regional studies was indeed duly tested by the unnerving results of contemporary synthesis. Australians seemed easily seduced when a lurking nationalism-imperialism surfaced with a vengeance in the interwar years. Boosters predicted a future national population of scores of millions, and even hundreds of millions. If a compact had been reached it was destabilized by inflated aspiration.

Outraged, the crusading pioneer geographer Griffith Taylor declared that severe environmental constraints would limit the total to approximately 20 million by the end of the 20th century. His forecast would prove stunningly accurate, of course, but at the time he was publicly reviled as a traitor and a croaking pessimist, and effectively blackballed. One of his simple textbooks was banned in Western Australia because of its emphasis on aridity and semi-aridity, and Taylor decided to leave for the United States at the end of the 1920s (Powell, 1993a). Twenty years later, the Snowy Scheme was welcomed as a timely re-assertion of dominion over nature, and of Australia's proud place in the world. 'Settlement' does not suit these convolutions if it is allowed to imply, as in other situations, a finished transaction. The replaying of the development card had shown again that Australia, like any other place, was a work in progress. So the transaction continues, as it must.

Acknowledgements

I am grateful to Tom Griffiths for encouraging and timely comments, and for permission to summarize some of his work at length.

References

Anderson, W. (2002), *The Cultivation of Whiteness: Science, Health and the Origins of European Australia*, Cambridge University Press, Cambridge.
Barrett, C. (1939), *Koonwarra: A Naturalist's Adventures in Australia*, Oxford University Press, Oxford.
Barrett, C. (1941), *Australia My Country*, Oxford University Press, Melbourne.
Barrett, C. (1942), *On the Wallaby: Quest and Adventure in Many Lands*, Robertson and Mullens, Melbourne.
Bonyhady, T. (2000), *The Colonial Earth*, Miegunyah Press, Melbourne.
Davison, G. (1989), 'The meaning of "heritage"', in G. Davison and C. McConville (eds), *A Heritage Handbook*, Allen and Unwin, Sydney.
De Courcy, C. (1995), *The Zoo Story*, Claremont Press, Melbourne.
Freestone, R. (1989), *Model Communities. The Garden City Movement in Australia*, Nelson, Melbourne.
Gascoigne, J. (1994), *Joseph Banks and the English Enlightenment. Useful Knowledge and Polite Culture*, Cambridge University Press, Cambridge.
Griffiths, T. (1996), *Hunters and Collectors. The Antiquarian Imagination in Australia*, Cambridge University Press, Cambridge.
Heathcote, R.L. (1965), *Back of Bourke. A Study of Land Appraisal and Settlement in Semi-arid Australia*, Melbourne University Press, Melbourne.
Haynes, R.D. (1998), *Seeking the Centre: The Australian Desert in Literature, Art and Film*, Cambridge University Press, Cambridge.
Hirst, J. (1983), *Convict Society and its Enemies*, Allen and Unwin, Sydney.
Inglis, K. (1974), *The Australian Colonists: An Exploration of Social History, 1788-1870*, Melbourne University Press, Melbourne.
Inglis, K. (1985), *The Rehearsal: Australians at War in the Sudan, 1885*, Rigby, Adelaide.
Inglis, K. (1987), 'Men, Women and War Memorials: Anzac Australia', *Daedalus*, vol. 116, pp. 35-59.

Jeans, D.N. (1988), 'The First World War Memorials in New South Wales: Centres of Meaning in the Landscape', *Australian Geographer*, vol. 19, pp. 259-67.

Jeans, D.N. and Spearritt, P. (1980), *The Open Air Museum. The Cultural Landscape of New South Wales*, Allen and Unwin, Sydney.

Macdonald, D. (1887), *Gum Boughs and Wattle Bloom, Gathered on Australian Hills and Plains*, Cassell, London.

MacCulloch, J. (1994), Creatures of Culture. The Animal Protection and Preservation Movements in Sydney, 1880-1930, unpublished Ph.D thesis, University of Sydney.

McGregor, R. (1997), *Imagined Destinies. Aboriginal Australians and the Doomed Race Theory, 1880-1939*, Melbourne University Press, Melbourne.

Martin, S. (1993), *A New Land. European Perceptions of Australia 1788-1850*, Allen and Unwin, Sydney.

Mazur, N.A. (2001), *After the Ark. Environmental Policy-making and the Zoo*, Melbourne University Press, Melbourne.

Meinig, D.W. (1962), On the Margins of the Good Earth: the South Australian Wheat Frontier, 1869-1884, Association of American Geographers, Chicago.

Powell, J.M. (1975), 'Conservation and Resource Management in Australia 1788-1860', in J.M. Powell and M. Williams (eds), *Australian Space Australian Time*, Oxford University Press, Melbourne, pp. 18-60.

Powell, J.M. (1976), *Environmental Management in Australia, 1788-1914. Guardians, Improvers and Profit*, Oxford University Press, Melbourne.

Powell, J.M. (1977), *Mirrors of the New World. Images and Image-makers in the Settlement Process*, Archon and Dawson, New Hamden and Folkestone.

Powell, J.M. (1982), 'Elwood Mead and California's State Colonies: An Episode in Australasian-American Contacts 1915-31', *Journal of the Royal Australian Historical Society*, vol. 67, pp. 328-53.

Powell, J.M. (1989), *Watering the Garden State. Water, Land and Community in Victoria, 1834-1988*, Allen and Unwin, Sydney.

Powell, J.M. (1991), *An Historical Geography of Modern Australia. The Restive Fringe*, Cambridge University Press, Cambridge.

Powell, J.M. (1991a), *Plains of Promise, Rivers of Destiny. Water Management and the Development of Queensland*, Booralong Press, Brisbane.

Powell, J.M. (1993), *'MDB'. The Emergence of Bioregionalism in the Murray-Darling Basin*, Murray-Darling Basin Commission, Canberra.

Powell, J.M. (1993a), 'Alfred William Howitt 1830-1908', Geographers. *Biobibliographical Studies*, vol. 15, pp. 51-60.

Powell, J.M. (1993b), *Griffith Taylor and Australia Unlimited, John Murtagh Macrossan Lecture for 1992*, University of Queensland Press, Brisbane.

Powell, J.M. (2003), 'The Empire Meets the New Deal: Encounters in Conservation and Regional Planning', *Paper Presented to the International Conference of Historical Geographers*, Auckland.

Powell, S.M. (2000), Mothering, Husbandry and the State: Conservation in the United States and Australia, 1912-1945, unpublished Ph.D thesis, Monash University.

Reynolds, H. (1989), *Dispossession. Black Australians and White Invaders*, Allen and Unwin, Sydney.

Reynolds, H. (1990), *The Other Side of the Frontier*, Penguin, Ringwood, Victoria.

Reynolds, H. (1992), *The Law of the Land*, Penguin, Ringwood, Victoria.

Robin, L. (2001), *The Flight of the Emu. A Hundred Years of Australian Ornithology 1901-2001*, Melbourne University Press, Melbourne.

Roe, M. (1984), *Nine Australian Progressives: Vitalism in Bourgeois Social Thought, 1890-1960*, University of Queensland Press, Brisbane.

Serle, G. (1973), *From Deserts the Prophets Come. The Creative Spirit in Australia, 1788-1972*, Heinemann, Melbourne.

Smith, B. (ed.) (1975), *Documents in Art and Taste in Australia: The Colonial Period, 1770-1914*, Oxford University Press, Melbourne.

Stubbs, B.J. (2001), 'From "Useless Brutes" to National Treasures: A Century of Evolving Attitudes Towards Native Fauna in New South Wales, 1860s to 1960s', *Environment and History*, vol. 7, pp. 23-56.

Trainor, L. (1994), *British Imperialism and Australian Nationalism. Manipulation, Conflict and Compromise in the Late Nineteenth Century*, Cambridge University Press, Cambridge.

Tyrrell, I. (1999), *True Gardens of the Gods. California-Australian Environmental Reform, 1860-1930*, University of California Press, Berkeley.

Ward, R. (1958), *The Australian Legend*, Melbourne University Press, Melbourne.

Williams, M. (1974), *The Making of the South Australian Landscape. A study in the Historical Geography of Australia*, Academic Press, London.

Chapter 8

Empire, Duty and Land:
Soldier Settlement in New Zealand
1915-1924

Michael Roche

Introduction

'Nothing has drawn the inhabitants of the Empire closer together than the Great War' declared French geographer Albert Demangeon (1925, p. 189), a now somewhat forgotten scholar who produced one of the more insightful outsider's accounts of the British Empire in the 1920s (Clout, 2004). In addition to the geographical expansion of the Empire he considered the question of what gave the Empire unity. Yet of the Dominions (Canada, Australia, South Africa and New Zealand), he observed 'the sacrifices entailed by the War and so loyally accepted have none the less given .. [them] .. a keener sense of their own interests' (Demangeon, 1925, p. 190). There is a sense in his writing of the tension between the metropole and the periphery of empire. Demangeon (1925, p. 275) was also conscious that great Empires of the past had gone into sudden decline and, for the British Empire, he offered some deductions about conditions that 'may tend to restrict its power or disturb its life'. His account also gave significant weight to the economic bonds of Empire and prefigures more recent interpretations which conceive of Empire as network (e.g. Ogborn, 2000). But if Demangeon was insightful about Empire, his commentary on New Zealand, as one of the most distant of the settler Dominions, was perfunctory and contained wholly within a discussion of 'Australasia' that dealt almost exclusively with Australia. Demangeon is interesting, however, in that he was writing contemporaneously with many of the events described in this chapter. Other geographers have subsequently focused attention on the British Empire. Christopher (1988) for example acknowledged Demangeon's early lead in his own study of the British Empire, at what he regarded as its maximum extent in 1931. Christopher focused on 'the overall unity of the Imperial experience and the influence which it brought to bear upon the landscapes of other continents' (1988, p. xi). More recently Lester (2002, p.25) has written of sites of colonization, including New Zealand, as 'nodes within an imperial network' through which 'flowed materials, people, and ideas'. The focus in this chapter is on ways in which changes took place within the Empire after it had reached its peak, specifically in New Zealand during and after World

War I, particularly in terms of the negotiated relationships between Empire and nation.

One of the ideas that coursed throughout the Imperial network during and after World War I was that of a discharged soldier settlement scheme. The idea provides the substantive context for this chapter. The Imperial dimension was provided by the visit of Sir Rider Haggard, probably now better remembered as a novelist, but in 1916 touring New Zealand on behalf of the Royal Colonial Institute in the guise of land reformer and advocate of Empire. Although the assisted emigration of British ex-soldiers did take place after the war it was not on a scale, or with the success, that was originally anticipated (Constantine, 1990; Fedorowich, 1995; Powell, 1980). New Zealand in 1915 was actually the first of the Dominions to pass legislation to resettle discharged soldiers. Similar schemes were subsequently initiated elsewhere in the Empire, including Kenya (Duder, 1993), Scotland (Leneman, 1989), and England itself (Lockwood, 1998). The intention in this chapter is to explore the distinctive political rhetoric of the soldier settlement scheme in New Zealand. It is less concerned with the legislative framework and the operational mechanics of the scheme than with using the Parliamentary debates on the principal and amending bills during and after World War I as a forum for discussion of the place of New Zealand within the Empire. It considers the degree to which the War hastened the development of a sense of national consciousness. Participation in World War I is usually taken to be one of the key markers in this particular journey (Phillips, 1989; 1990; 2000; Sinclair, 1986). New Zealand, with a total population of 1.2 million, contributed 100,000 men to the war effort of whom 16,000 died of wounds and disease. A number of war memoirs express the view that New Zealand entered the war as part of the Empire but emerged from it as a distinct nation after five years of horrific fighting, but much of the fine detail of this transformation is still missing.

In a review of 20th century war histories in New Zealand, Deborah Montgomerie (2003, p. 62) suggests that 'War, we are told, brought us to self-consciousness as a nation, but the details of the process (particularly the way it worked itself out in the messy periods between the wars) remain indistinct'. Separately Scott Worthy, in considering New Zealand war memoirs, remarks that 'Equally divided was discussion over whether war brought empire closer together or whether it created a sense of separate New Zealand identity' (2004, p. 23). Written remembrance was one way in which the experiences of the war and its impact on national identity were played out (e.g. Burton, 1935). Public observances at Anzac Day were another (Sharp, 1981). Memorials of various types were a further element (Pawson, 2004), while a fourth arena is to be found in the responses of the state which included repatriation, pensions, hospital care, and most notably a land settlement scheme for would-be farmers (Powell, 1971).

The perspectives of Demangeon, Montgomerie, and Worthy provide the spring-board for this chapter, which looks at the *Discharged Soldier Settlement Act, 1915* and its six amendments through to 1924. This legislation provided a forum in which Empire and duty were discussed from the second year of the war through to the mid 1920s, at a time when the New Zealand rural economy was struggling because of fluctuating export prices on the British market.

In 1935 W.P. Morrell, an expatriate New Zealand historian then on the staff at the University of London, published *New Zealand,* a volume in the series 'The Modern World, A Survey of Historical Forces', edited by H.A.L. Fisher. Fisher also contributed a foreword that contended:

> By curious paradox this people which more than any other body of British emigrants or descendants of British emigrants still retains the characteristic of the Old Country, has come to feel itself a separate nation by reason of the very circumstances which it has been called upon to play in the general life of the British Empire (Fisher, 1935, p. viii).

He continued to allude to New Zealand's role in the South African War and World War I which he saw as exemplary 'moral preparation' for the 'closer association between the free and equal communities which now constitute the British Commonwealth of Nations' (Fisher, 1935, p. ix). This effectively summarized Morrell's approach to writing a history of New Zealand. For him, World War I was crucial to the development of a new sense of nationality. He suggested that, 'the men who fought for the Empire, also fought for New Zealand. The Empire belonged to the realm of imagination; New Zealand belonged to the realm of experience. They thought of it as their own country' (Morrell, 1935, p. 112). This did not, however, stop him from being critical of the government policies that promoted the settlement of discharged soldiers onto the land. Morrell's concerns are drawn from those advanced by New Zealand economic historian J.B. Condliffe (1930, p. 224), that 'in effect the Government turned loose in the real estate market 22,585 new purchasers armed with £22,627,864 of borrowed money' with resultant rampant inflation in land prices.

Discharged Soldier Settlement in New Zealand 1915-1924

In common with Australia and Canada, the soldier settlement scheme that operated in New Zealand offered numerous individuals the opportunity to farm (Fedorowich, 1995; Powell, 1980). In New Zealand, some 10,000 discharged soldiers were settled on the land between 1916 and the mid 1920s. The scheme has attracted scholarly attention mainly in terms of discussions about the uneconomic size of farms, their location on poor quality bush lands, their remoteness, the relative lack of capital and experience of the settlers, and the overall impact on land prices in the early 1920s (Gould, 1992, 1994, 2000; Roche, 2002).

The empirical focus of this chapter is on the Parliamentary debates, especially the second readings of the Discharged Soldier Settlement Bills. Table 6.1 indicates the number of speakers and pages that each of the second readings occupies in the *New Zealand Parliamentary Debates* and provides a rough index of their political significance. It is notable that the bills attracted more attention as time passed and that relative to the House of Representatives (the upper house), the Legislative Council had little to say. Only a fraction of each debate was given over to discussions of Empire, duty and nation. Much was, however, taken for granted and the exchanges usually concentrated on the specific provisions of the original bill and its amendments.

**Table 6.1 Number of Pages of Parliamentary Debates for Discharged Soldier
          Settlement Acts 1915-1924**

Second Reading of Bill	*1915*	*1916*	*1917*	*1919*	*1921-2*	*1923*	*1924*
House of Representatives[1]							
Number of Pages	18	33	36	34	32	88	3
Number of speakers[2]	14	9	13	20	14	27	5
Legislative Council							
Number of Pages	1	12	4[3]	12	1	11	4
Number of speakers	1	9	5	6	0	6	4

Source: New Zealand Parliamentary Debates (hereafter NZPD) 1915-1924.

Notes

1 There were 76 MPs in the House of Representatives during this period. In 1915, 54 of
   these would have been classed as rural, 18 from cities and the remaining four from
   provincial towns.
2 Interjections are not counted. The New Zealand Parliamentary Debates lists an MP's
   electorate when they speak but not when they interject. This means that George Witty, a
   Liberal MP, who peppered Massey's introduction of the 1915 Bill with interjections is
   not included.
3 There were however ten pages of First Reading debate in Legislative Council in 1917.

Discharged soldier settlement was initially debated by only a small number of
MPs predominantly holding rural electorates. Consistent participants were those
from the bush frontier electorates of the central North Island. There are, however,
some other cleavages; for instance between largely South Island Liberal MPs and
North Island Reform (i.e. conservative) MPs, a division that continued from 1915
into 1916. North Island Reform MPs numerically dominated the debate of 1917; a
situation that was reversed in 1919 after the coalition National government broke
up. South Island MPs were also more vocal in this discussion. In the 1921-22
amendment bill debate North Island government Reform party MPs predominated.
The 1923 bill represented a major response by the government to the problems of
meeting rents when primary produce prices had collapsed, and MPs from many
rural electorates spoke. In comparizon the 1924 Bill was narrower in scope and
attracted minimal discussion. This changing geography of support for soldier
settlement of itself does not reveal the detail of the discussion of Empire and
nation. It foreshadows, however, some commonality of opinion amongst rural MPs
who articulated the stances of their electorates about soldier settlement and in this
context, views about Empire, duty and land. This situation was reinforced by the
age of many of the MPs, particularly those born before 1880, a number in Britain
or Ireland, who had lived much of their lives at a high tide mark of Victorian
imperialism.

The Discharged Soldier Settlement Act 1915

In introducing a bill to facilitate the post war settlement of discharged soldiers in 1915 Prime Minister Bill Massey was continuing with a tradition of closer settlement that stretched back a generation in New Zealand and was grounded in an agrarian myth (Arnold, 1981; Brooking, 1996). The bill was introduced only weeks after a coalition 'National Government' was reluctantly formed by Reform and Liberal MPs as a response to war-time conditions. It was agreed that only necessary legislation would be brought forward for the duration of the war. Massey's introduction of the bill was relatively low key; he suggested that the title of the bill described its purpose, spoke of various blocks of available crown land adjacent to Wellington, the capital city, and about the possibilities of resettling some men on apple orchards.[1] Subsequent speakers expounded on the importance of duty to Empire and country. Edward Newman reiterated that 'the duty of the country to the soldiers who are fighting our battles at the front is not ended by a war pension' (NZPD, 1915, 174, p. 213). William Jennings attributed more spiritual significance to the link between duty and the land in proclaiming, 'The land that is given to the men who are fighting the battles of the Empire should be almost inalienable; it should be for them and for their people hereafter' (NZPD, 1915, 174, p. 215). The importance of duty was reiterated by the Attorney General, who reminded his listeners, that 'these men have gone out of New Zealand to fight for us, and we have a duty to them – a duty which I am sure parliament and the people of this country will eagerly and earnestly discharge' (NZPD, 1915, 174, p. 218). Empire and nation were mutually constituted in the minds of many of the MPs. These sentiments were expressed in most stirring terms by former Auckland city Mayor James Parr, the Reform MP for Eden, who having put behind him issues of public utilities, slum clearance, town planning and local body amalgamation, spoke evocatively of picturing 'in the future all over our country colonies of these returned soldiers planted, as the ancient Romans did, in different parts of their great Empire after their victories. These soldier settlements in ancient Rome lasted for hundreds of years holding the soldier tradition, possessing the patriot traditions of Rome, just as we hold that Britain shall ever be imperial, free, mighty. .... Here will be nurtured the patriot soldier sentiment combined with our democratic citizen ideals; and all over New Zealand we shall have these men settled down as the finest citizens in the land, because they were not afraid in that great hour of stress and trial to go forth and do their duty like men' (NZPD, 1915, 174, p. 218). This was also one of the best examples of a heroic conception of warfare being projected forward to a similarly heroic view of land settlement.

There were some subsidiary threads to the discussion whereby the soldier settlement scheme was seen as a corrective to the evils of the drift of population to the city (Webb, NZPD, 1915, 174, p. 218). It was also argued that its provisions should apply equally to nurses who might wish to be involved in poultry farming or orcharding (E. Newman, NZPD, 1915, 174, p. 214). Undertones of class were also manifest when James McCombs (Labour MP for Lyttelton) declared that the proposed legislation did not go far enough and that 'it savours of class legislation – legislation in the interests of those troopers – and they are less than 20 per cent –

who have come from the country districts'. He asked 'what of the industrial workers?'(NZPD, 1915, 174, p. 226).[2] Phillips (1987, p.305) from a 1:20 sample of soldiers estimated that farmers did indeed constitute 20.1 per cent of the main body of troops.

The Discharged Soldier Settlement Act 1916

The original Act was amended in 1916 when the eligibility criteria and financial assistance was extended. Conscription was also introduced in that same year. Massey in his capacity as Minister of Lands rather than Prime Minister introduced the second reading of the bill to the House of Representatives, voicing his belief that it would necessitate only a short debate. By 1916, with the casualty lists growing, thoughts of a short and heroic war had given way to a grimmer reality. Massey's tone and that of other speakers was, however, noticeably more ideological than in 1915. It was, he declared, 'the duty of the State to do everything it possibly can do to provide lands for those of our soldiers – those who are fighting for us now or who have been doing so in the past – and assist them to get onto that land in any way we can' (NZPD, 1916, 175, p. 841). Subsequent speakers reiterated this view. Massey then followed this with the declaration that it was 'our duty on account of the country and on account of the Empire to assist returning soldiers onto farms' (NZPD, 1916, 175, p. 842). At this juncture he linked his words to the visit of novelist Sir Rider Haggard, whom he described as 'that well-known Imperialist' (NZPD, 1916, 175, p. 842). Haggard was touring New Zealand as part of a Royal Colonial Institute deputation for the post-war resettlement of discharged British soldiers throughout the Empire (Constantine, 1990). Massey went so far as to read to the House his letter to Haggard in which he exclaimed 'I think it is of the utmost importance to the Empire that after the war any immigration which takes place shall be directed to the overseas dominions rather than be allowed to find a way to alien lands' (NZPD, 1916, 175, p. 843). But having made this pledge to Empire he then offered a significant qualification, to the effect that good available land was limited in New Zealand and that the New Zealand soldiers had 'first claim'. This commitment to New Zealand soldiers he repeated to the House. It also underlay his desire to allow New Zealanders who had served with the Imperial or Australian forces to be included within the extended scope of the legislation.[3] Haggard separately reported Massey as 'not very definite but full of goodwill as regards land settlement and employment' (Haggard in Dunne, 1993, 17).

From other members Haggard's mission incurred scathing criticism. Liberal MP John Hornsby suggested 'let him tell the Landlords of England, Scotland and Ireland that they should make room for the soldier who is going back. Why should he come to New Zealand or Australia, or the Cape and ask that room be made for men who have fought the battles of Empire' (NZPD, 1916, 175, p. 849). Not only did Hornsby see it as a matter of duty to make provisions for 'our own returned New Zealanders' (NZPD, 1916, 175, p. 849) but he construed this in terms of an agrarian myth stating 'let us put the men on the soil of this country. The men on the soil are the mainstay ... Make provision for putting the people upon the land and thus secure to New Zealand the soundness and success of its future' (NZPD, 1916, 175, p. 852). Hornsby's condemnation of landlords was a reflection of his political views; he was a supporter of closer land settlement and opposed to land aggregation (Hamer, 1996).

But this ought to be balanced against others who spoke in terms of Empire. Sixty-four year old David Buddo (Liberal MP for Kaiapoi) a former North Canterbury Mounted Rifles Volunteer (1902-03) remarked that 'those men will have assisted to keep New Zealand free and still part of the Empire to which we belong, and it is our duty to see that generous treatment is afforded to them' (NZPD, 1916, 175, p. 863). Others such as Alexander Harris spoke in heroic terms of 'the men who are so nobly fighting the battles of Empire at the present time' (NZPD, 1916, 175, p. 866), while John Anstey (Liberal MP Waitaki) still saw the New Zealand soldiers as 'fighting for the Empire' (NZPD, 1916, 175, p. 870).

Other MPs spoke of a threat to New Zealand, with William Veitch even making the claim that 'if the nation had not arisen to the occasion and taken part in this great conflict – New Zealand would be a German possession today' (NZPD, 1916, 175, p. 855). Veitch also returned to older political debates by identifying the large estate owners as a barrier to closer settlement. There followed a discussion of the soldier settlements established in the 1850s and 1860s, as well as reflection on the parallel events in the United Kingdom, where the report of Sir Harry Verney's Committee on the settlement of discharged soldiers on the land in England and Wales had recently been released (Miller and Roche, 2003).

The Discharged Soldier Settlement Act 1917

The 1917 Bill was intended to make additional land available more quickly by ex-soldier initiated purchase by the Crown of freehold land, and by loans for the direct purchase of freehold lands by discharged soldiers. Massey again opened the debate with the gambit that that he did not believe there would be much difference of opinion about the bill. After outlining the details of the amendment he paused to laud the efforts of 'patriotic' men throughout the Dominion who had assisted returned soldiers to become farmers. With the war now in its fourth year the tenor of the debate had again changed.

Soldier settlement was held to be still 'of the very greatest national importance to the future of the country' (NZPD, 1917, 180, p. 49), but the concerns were now rather different. They crystallised around the attractiveness of farming, the evils of urban drift, the quality of land, and escalating land prices. The heroic view of war still appeared to inform the politicians who spoke of physically hardened heroes returning so that the fight would continue, but it would now take the form of a fight against nature to convert the undeveloped forestlands of the country into productive farmland. Thus David Buick MP for Palmerston spoke confidently of the future; 'many of the men now returning are somewhat broken in health, while those who come back after the war will be strong and robust and fit to go into the hard work of farming' (NZPD 1917, 180, p. 51).[4] This theme was later taken up by Christchurch East MP and city mayor Dr Henry Thacker, a political opponent of Massey, who proclaimed 'Anybody would think that the soldiers who are coming back – those grand men who are practically winning the war – were in their second babyhood and wanted sucking-bottles. Nothing of the sort. The men are coming back virile, and strong, and keen; they will infuse new blood into the dormant blood of the Dominion' (NZPD, 1917, 181, p. 96). Allied to this was a belief that 'groups of men who had been together at the war would naturally want to settle down together' (Newman NZPD, 1917, 180, p. 41), and that battle-hardened soldiers, formerly town dwellers, would now seek to take up farms. Richard Hudson, who had actually had a military caree, retiring as a Captain in the Royal Artillery, commented 'the men who went from the cities in many case may not be inclined to go back to their old occupations in offices and shops' (NZPD, 1917, 181, p.94). Labour MP Paddy Webb, cautioned that those who had been wounded would find difficulties in returning to their old occupations.

The unshakable view was that expressed by Mr Jennings 'we must look to the land that we have, to look for our national prosperity in the future' (NZPD, 1917, 180, p. 49). In this context the complaints of other rural MPs such as Newman about the city and city life gain additional significance, 'Surely these men are better in the country than crowding into the cities. In my opinion, we want to get people away from the cities and into the country, so long as they are likely to make good settlers. It is a great misfortune of this Dominion that so much of the country population has been drifting into the cities' (NZPD, 1917, 180, p. 42). Webb alone opposed this viewpoint when he injected a note of sarcasm in suggesting that 'the only chance they have to go into the country is by putting a swag on their backs and tramping there. They cannot buy land in the country at the price now asked for it' (NZPD 1917, 180, p. 57). His more serious concern, however, was the longer term implications of land purchases on a rising market. His unpalatable message was that 'if the State has to pay too high a price for the land it will mean the bankruptcy of many of the returned soldiers who settle on it' (NZPD 1917, 180, p. 59). Specific discussion centred on land purchased at £65 per acre for what became the Kopane Soldier Settlement (Roche, 2002).[5] In 1917 Webb was jailed for opposing conscription and other Labour MPs were jailed for sedition.

The wider anxieties differentiated the New Zealand population into British and 'Other', as for instance when the MP for Egmont claimed that, 'for some time there has been much feeling aroused in my district over the purchasing of considerable quantities of land by aliens – land that might otherwise be available for soldiers. They are not enemy aliens, however, some of them may not be naturalized … but there is a principle involved, and it is that the land of New Zealand should be available for men of British birth in preference to foreigners, however desirable' (NZPD 1917, 180, p. 61). This view reveals the extent to which land was to be centrally associated with emerging conceptions of nationhood.

As on previous occasions the Legislative Council debate was inconsequential though there was some discussion of an earlier round of soldier settlements in the 1860s.

The Discharged Soldier Settlement Act 1919

This amendment broadened the eligibility criteria to include widows of ex-soldiers and those who served in the armed forces but were not posted overseas. Familiar sentiments included the view that each soldier had been away 'fighting for his country and the Empire' (NZPD, 1919, 185, p. 304); with that came the obligation that the country should place them on the land. Farmer MP George Forbes, a future Prime Minister, interpreted this to mean 'the right to be settled on good land instead of being banished to the mountain tops' (NZPD, 1919, 185, p. 288). Patriotism was an undercurrent to the discussion especially as it affected the prices being sought for land. Christchurch city Liberal MP Harry Ell claimed it was 'a very disloyal and unpatriotic thing for any body of men in this country to get in and buy a block of land knowing perfectly well that the Government are out to buy that land for soldiers' (NZPD, 1919, 185, p. 296). Richard McCallum was more critical of land speculators describing them as having no sense of shame and suggesting that some at least, being non-British, were not eligible for conscription: 'Without family representation at the fighting-front they have been buying up land for speculation purposes during the absence of likely buyers who have done their highest duty to the State' (NZPD, 1919, 185, p. 298).

The Legislative Councillors were also more outspoken in lamenting land speculation and rising land prices. John Paul (MLC for Otago) linked both of these concerns together in remarking 'whereas the soldier by his going-away and fighting for this country made the position of the landowner secure, the least every landowner can do is to dispose of, at a fair price to the soldier, any land that he may need' (NZPD, 1919, 185, p. 193). Paul had a trade union and Labour party background. He had been an appointee to the Legislative council in 1907. Unlike some in the Labour movement he supported New Zealand entry into World War I having been involved in defence and patriotic organizations before and after the war (Olsson, 1996).

The Discharged Soldier Settlement Acts 1921-22, 1923 and 1924

The three final amendments to the Discharged Soldier Settlement Act can be dealt with together. Even though the 1923 amendment was the most debated of all the Bills, the time spent talking about Empire, nation and patriotism was much reduced compared to earlier amendments.

The duty to the soldiers was again raised by McCallum in the 1921-22 amendment that was largely concerned with enabling the postponement of interest payments in response to falling export prices. The troops had been on the battlefields of France 'fighting for the Empire' George Mitchell reminded the House (NZPD, 1922, 193, p. 830). New Zealand-born Gordon Coates, who served as a Major in the New Zealand Expeditionary Force and was a rising politician (PM from 1925-1928), used the occasion to recollect that 'Everyman who came back to this country from the war, every person in New Zealand and I believe, every honourable member in this house advocated and urged the Government to settle the soldiers on the land' (NZPD, 1922, 193, p. 823). But it was Mitchell who injected a memorable phrase into the debate when, in questioning the loyalty of those who sold land to the government at inflated prices, he claimed of one unnamed individual that 'his patriotism was all in his pocket' (NZPD, 1922, 193, p. 829).

New Zealand farm export prices in Britain, the major export market, had slumped in 1921. The knock-on effect had been realized as early as 1919 by some politicians. With land rentals pegged to an unrealistically high war-time valuation, many soldier farmers quickly went into debt. By 1923 an amendment was introduced to establish a Dominion Revaluation Board to which ex-soldiers could apply for rent and mortgage reductions and relief of arrears. Defending the scheme, Massey reiterated the idea of duty, 'we have got to protect the soldier settler as far as we possibly can' (NZPD, 1923, 201, p. 631). The soldiers had met their duty to the state through active service. Now the state had to do its duty by the soldiers. Massey's political opponents would have also read into this a commitment to protect his rural voting constituency. Labour's Frank Langstone did allow, however, that 'the soldiers on the land ought to have a chance of making a living. Today they have not had that chance' (NZPD, 1923, 201, p. 666).

A number of speakers affirmed Massey's comments about duty to the soldiers (e.g. McKay, 1923, 201, p. 696). It was probably expressed in the most heroic terms by Mr Burnett, a stalwart supporter of rural society against urban interests, when he reminded the House that 'romance is not dead among us yet, and we felt and rightly so, that for these men who had gone to the front to fight for us and the Empire, putting aside for the time being one's life ambitions and everything else, we could not do enough to show our appreciation of them' (NZPD, 1923, 201, p. 678).

Members of the Legislative Council spoke more briefly but, following the lead of Sir Francis Bell, reflected on the duty of the government. Bell, a former lawyer usually not given to hyperbole, spoke in fulsome terms of the scheme: 'I suppose we may safely say that no part of the British Empire – indeed, that no country in the world – had done so much for its returned men as had been done by the Dominion of New Zealand. That is the undoubted fact' (NZPD, 1923, 202, p. 5).

The final amendment to the principal Act took place in 1924 and was intended to extend the time available for soldiers to seek rent and mortgage relief. The debate was short and completely focused on the technical aspects of the Bill. There were no appeals to duty, nation, patriotism or Empire.

Soldier Settlement as the new Battle Front

Overt utterances about Empire, duty and nation comprized a comparatively small part of each debate on discharged soldier settlement legislation. Reading them, there is a sense of the 'talking up' of empire and nation, as successive speakers endeavoured to show that their loyalty was above reproach. Ulster-born Massey's support for Empire was well known and is summed up best by Gould (2000a, p. 313), who quotes him to the effect that 'if it were possible for the point of view of

New Zealand and the point of view of the Empire as a whole to come into conflict, I would go for the Empire at once'. Although Massey's response to Rider Haggard's soldier settlement scheme appeared contradictory, he nevertheless articulated the view that New Zealand was unquestionably part of the British Empire.

The imperial connection was also strongly felt by some New Zealand-born MPs. Thus the enthusiasm of lawyer James Parr (who had likened soldier settlement to the soldier colonies of ancient Rome) for the imperial cause extended to support for the Empire Trade League to keep German goods out of New Zealand. He 'exhibited a deep interest in Imperial affairs' (Anon, 1926, p. 659) as New Zealand High Commissioner in London. Welcomed as High Commissioner by the Royal Colonial Institute in 1926 he declared 'I firmly believe the best cement of Empire today is mutually profitable trade' and lamented the inroads that American motor vehicles and cinema were making in New Zealand (Parr, 1926, p. 603). While his membership of the Imperial Defence Committee (1928-30), the Empire Marketing Board (1927-30) and the Imperial Shipping Committee (1926-30) may be regarded as official duties, Parr's personal commitment to Empire was evident in his support for, and lectures to, Lord Beaverbrooke's Empire Crusade in 1931 (Taylor, 1972). In the late 1930s he was a Vice President of the Royal Empire Society.

Gordon Coates' biographer describes his willingness to enlist for active service at the age of 38 when an MP, married and with a family, as stemming from his support of empire but also, not insignificantly, from a sense an excitement and adventure (Bassett, 1995, pp. 47-48). Others, such as Edward Newman, helped in 1920 to establish a trust fund from New Zealand sheep owners' wartime wool profits to provide for relief of British Seamen and to assist widows and orphans, most notably through farm training schemes in New Zealand (Goodall, 1962). Newman saw these sailors as having served in the defence of the Empire. Yet some MPs such as Thomas Sidey, who in obituaries was described as 'passionately devoted to the Mother Country', did not articulate similar sentiments during the debates on the Discharged Soldier Settlement Bills (Anon, 1975, p. 172).

Opposition MPs were faced with a difficult situation, one that was not entirely diminished by the formation of a coalition National government in 1915. Soldier settlement was not directly opposed during the debates in question by any MP but there was oblique criticism during the war. This criticism came most consistently from a group of MPs who would eventually coalesce to form the Labour Party. Thus Webb, Veitch, and McCombs contested some of the objectives of soldier settlement. In contrast, John Paul, a union-affiliated member of the Legislative Council, supported the war and also wholeheartedly supported the soldier settlement schemes.

Empire and nation were elided in the politicians' speeches. The troops from New Zealand were fighting for the nation and for the Empire. Empire was also the label under which other soldiers might be admitted to rural New Zealand. These men had served under the same flag and would thus also be welcomed as potential farm settlers. That New Zealanders might leave for other countries outside the Empire, even to allies such as the USA, was regarded as a matter of concern for the future vitality of the Empire. Even Massey, a clear supporter of Empire, had been obliged to choose his words carefully in responding to Haggard's proposal for the post-war resettlement of British soldiers in New Zealand. While affirming New Zealand's loyalty to the Empire, he had to insist that New Zealand troops must have preference and that there would be little enough land available in any case. The interconnected nature of Empire and nation in the debates over the discharged soldier settlement bills does not show the same sharp cleavages that Worthy (2004) reveals in the literature of World War I remembrance, where Gallipoli and the war are seen as forging a distinctive sense of New Zealand nationalism.

The heroic view of warfare that carried through in MPs' minds to 1916 saw the returned soldiers as physically hardened by their trials and triumphs, and ready to return to New Zealand as stronger men, keen to take on the task of farming and making new farms. Having fought together there was a view that ex-soldiers would want to farm together on adjacent properties. While it was conceded by some that these new farmers deserved more than the bare necessities of the pioneers, others envisaged the farm as the basic social unit, the returned soldier having taken a wife and started a family. Another recurring thread of moral concern about the returning soldiers was articulated by Jennings, who cautioned that 'we do not want any idlers among the returned soldiers, and there is nothing better than farm work for them to do' (NZPD, 1917, 181, p. 51). A particular concern was that the discharged solider would languish in town, avoiding hard work and responsibility and become a 'shirker'. The routine and virtues of hard farm work were seen as an ideal cure. In contrast to the morality and order of the farm landscape, in the towns the returned soldier would be subject to the temptations of idleness and other urban vices (Fairburn, 1975).

Several groups were marginalized in the discussions about nation that surrounded the Discharged Soldier Settlement Bills. These included German-speaking residents, other non-British settlers, Maori and women. As the war continued those people of non-British ancestry were increasingly ostracized. There was considerable anti-German sentiment by 1915 (King, 1998).[6] People of German descent were positioned as 'enemy aliens' and a threat to Empire and nation. However, this did not mean merely interning German nationals in New Zealand for the duration of the war. In 1917 Charles Poole, the MP for Auckland West, argued that the 280 German prisoners interned on Somes Island ought to be put to work 'to break in the virgin forest to have it prepared for the returned soldiers' (NZPD, 1917, 181, p. 91). In the same year another MP claimed an 'unnaturalized German' was purchasing farms in the vicinity of Palmerston North and that this was neither a patriotic nor a productive use of land. Guthrie, the Minister of Lands undercut him by asserting that the man in question claimed to be Swiss (and though presumably a German speaker was not a national).

The loyalty of other settlers of non-British origin was also called into question, particularly the Croatians from Dalmatia who worked on the North Auckland gumfields (Trlin, 1979).[7] Poole went so far as to argue that the regiment that they were organizing, which he referred to as 'Jugoslavs', ought, 'if we want to test their loyalty and utilize them to best advantage', to be given the task of developing forests into farm land for the returning soldiers (NZPD, 1917, 181, p. 91). As hostilities continued and the first discharged servicemen took up land under the Soldier Settlement Act, non-British residents were further targeted as 'aliens'. Some of the more zealous MPs suggested that their land ought to be confiscated to provide farms for the returning soldiers. In the heated discussions about the impact of rising land prices on the settlement scheme, war as well as land aggregation by 'aliens' was specifically targeted. The latter was regarded as an unpatriotic act when undertaken by New Zealanders but was particularly loathed when carried out by 'aliens'.

Maori appear in the discussion about soldier settlement in several ways. They are written out of the discussions of 19th-century soldier settlement schemes in New Zealand that formed part of the debate on the 1916 Bill. Jennings, for instance, launched into a long discussion of the success of these settlements and the contributions to public life in New Zealand made by descendants of the original settlers. He made no mention of the fact that some of the settlements he was referring to, such as Howick, were established in the 1850s as ring of military settlements designed to protect Auckland from attack by Maori, or that Hamilton and Cambridge were part of a defensive curtain of redoubts erected in order to seize and hold the Waikato after the 1863 invasion. Some 3000 former militia were settled on land that the colonial government confiscated in the aftermath of the Land Wars. By 1916, the imperial character of this earlier episode of land settlement had been lost from mainstream political memory.[8]

It was quickly realised that a considerable amount of land would be required for the returning soldiers. Since suitable Crown land was limited in extent, most would have to come from the freehold estate and Maori land. Initially, the difficulty in securing land was seen in terms of the reluctance of the remaining, albeit much reduced, great estate owners to sell their land. They had been a particular target of Liberal Government rhetoric in the 1890s (Brooking, 1996). Massey had acknowledged that the government was also buying Maori Land for soldier settlement as and when it had 'the opportunity' (NZPD, 1916, 175, p. 841). The Maori land question came to the fore again in the Discharged Soldiers Settlement Loans Bill of 1919. After another MP alluded to the 'hundreds of thousands of acres of virgin land that are now lying idle in the Urewera country' (NZPD, 1919, 184, p. 121), Mr George Russell, the MP for the Christchurch seat of Avon, proposed that the government should purchase this Maori land for returning soldiers. Apirana Ngata, the MP for Eastern Maori, entered the debate in response to this suggestion. He addressed the government bluntly: 'when ever there is any discussion about the unoccupied lands of the Dominion ...... the thoughts of the majority of the members of this house fly most naturally and easily to Maori land' (NZPD, 1919, 184, p. 142). Ngata then identified specific blocks of Maori land that had been sold cheaply to become soldier settlements, before outlining his own plan for the repatriation of Maori returned soldiers. He then played his trump card, the exemplary performance of Maori soldiers during World War I (Pugsley, 1995). Ngata was able to assert that 'the Maori Pioneer Battalion had established in France a splendid reputation for its work in the war' (NZPD, 1919, 184, p. 144). While this was not disputed, it nevertheless remained the case that less than two per cent of Maori soldiers acquired land under the Discharged Soldier Settlement Act, compared to about ten per cent of Pakeha soldiers (Gould, 2000b, p. 299).

Women were referred to on more than one occasion in their own right, as well as, later on, as wives. Wives and Empire were linked by Anstey, who claimed that 'widows should stand in the shoes of their dead husbands who have died in the defence of Empire' (NZPD, 1919, 185, p.291). Wilford, the Liberal leader in 1923, also urged MPs to recognize the role of the wives who had 'done so much to help their husbands in the re-establishment of homes after strenuous war service' (NZPD, 1923, 201, p. 712). Although by 1923-24 the discussion concentrated on men as ex-soldier farmers, the question remains of whether women were really 'written out' of the farm landscape, having only been 'written in' during 1915-17 as parliamentary speakers tried to out do each other in patriotism. Nevertheless, some women did apply under the Act, for example Nurse Schaw, who briefly ran a poultry farm on Cloverlea Soldier settlement.[9]

Conclusion

Almost without exception the MPs who spoke on soldier settlement considered New Zealand to be a nation. Arguably, the war accelerated this perception. But as Sinclair (1986) has observed, they also spoke of the nation as part of the Empire. The New Zealand Parliament was predominantly a gathering of rural interests and the urban elite, and as such articulated views that were rooted in 19th century visions of Empire. Debate was concentrated largely within ranks of rural MPs, often taking the form of sometimes subtle contests between Reform and Liberal MPs.

Much of the discussion of nation and Empire was anchored in the idea of 'duty', although in fact this had several layers of meaning. There was New Zealand's duty to Empire, exemplified by the armed forces raised for service in the war. There was also the duty of the New Zealand state to provide for the needs of its returning soldiers. Empire and duty tended to coexist in the politicians' utterances, particularly during the course of the war. Even afterwards they were perhaps more likely to see Empire and nation as mutually constituted than the writers discussed by Worthy (2004), who had actually seen active service. Empire was invoked to explain New Zealand's participation in the war but the idea of a national duty to the returned soldiers was invoked to justify the discharged soldier settlement scheme. Initially at least, some MPs saw soldier settlement as an extension of an heroic model of warfare transposed from one battlefield to another: land settlement, where the fight was against nature. As conditions became more difficult for all New Zealand farmers in the 1920s, the idea of duty remained but the focus on Empire tended to recede in the debates.

In these discussions of nation, those of non-British stock, like the Croatians and other 'aliens', were excluded from the inner circle. Likewise, the Maori posed a similar difficulty for MPs. The war-time contribution of the Maori Pioneer Battalion could not be faulted, but the desire to alienate remaining Maori land for soldier settlement suggested that the emerging nation was still very much a 'white' derivative of Britain, although in some ways proclaiming itself superior to the 'Old Country'. The emerging national self consciousness was also still strongly influenced by a sense of its pioneering rural past, even though by 1921, 49 per cent of the country's population of over 1.2 million lived in towns of more than 1000 people.

By using the Discharged Soldier Settlement Bills as a frame through which to discuss politicians' attitudes to Empire, duty and nation, it is possible to identify some threads that do not appear in other work on these themes, particularly in connection with land. These debates also provide the means of tracking imperial sentiment for several years after the war had ended. The views of politicians are not necessarily typical of the wider population, although in this instance, many of the rural MPs appear to have been articulating at least some of the commonplace opinions of their constituents. This bears on Worthy's point over whether World War I accentuated imperial feeling in New Zealand or aided the development of its national identity. In these debates on soldier settlement, a sense of duty to Empire remained strong, even in the post-war era, particularly but not exclusively amongst older MPs. However, this view was most frequently expressed in terms of the co-existence of the Empire and the New Zealand nation. The exclusive nature of this view of nationhood is also clearly apparent in the debates over the Soldier Settlement Bills, particularly in the exchanges over enemy aliens, Maori and women. To oppose the Discharged Soldier Settlement legislation was to oppose the government and the war effort, which some Labour MPs did to their cost.

Land was important to the construction of ideas of Empire and nation in New Zealand, as it was in the emergence of nationalism in both 19th-century Europe and other former colonies (Sinclair, 1986). In New Zealand, the issue was conceived of in terms of the potential threat to the nation and Empire posed by land aggregation by 'alien' landowners. Similarly, the sale by New Zealanders of freehold land to the Crown for soldier settlement at inflated prices was also regarded as unpatriotic. Indeed, after World War I, 'land' was one of the ways in which pre-war party politics surfaced in parliamentary discussions about soldier settlement, particularly in regard to the further break-up of the great estates. The place of Maori land in these discussions was also revealing. Its continued purchase would ease the problem that Massey foresaw in providing sufficient Crown land for soldier settlement, but was in itself a continuation of a much longer political and economic contest (see Brooking, 1996a). Maori MPs pointed to the record of Maori troops during the war, and argued that participation in the conflict had admitted them to full citizenship (Bennett, 2001; Phillips, 1990). But the situation was more complex than the soldier settlement debates suggest. There were assimilationist overtones in some of the discussions, which constituted a marked point of difference with post-war Australian nationalism (Meaney, 2003).

What is particularly interesting in terms of soldier settlement is the way in which the land issue permeated the political debate. Land and duty were linked, typically by rural MPs. There was a duty to defend the land of the Empire and of New Zealand as a nation. The state's reciprocal duty to the soldiers was to provide land for farms. This land was presented as a new front where a further battle against the forest wastelands could be fought by the hardened men who had returned from the war. For those who were too badly wounded to participate in this contest, fruit orchards were offered as an ideal way to recuperate physically and mentally from the war. This unpacks at least one part of the process of increasing national self consciousness alluded to by Mongomerie (2003). However, a straight forward sequencing, whereby the sense of 19th century identity as colonials gave way to one in which Empire and the New Zealand nation were mutually constituted in the early 20th century, and to a sense of separate nationhood after 1918, is too simplistic. The emphatic view of Demangeon, perhaps too much swayed by the temporal closeness of events, that the war brought the Empire closer together, can be challenged. Likewise Christopher's focus on the material shape of Empire and the outward similarity of imperial institutions tends to underplay the growth of national sentiment in the Dominions. Yet Demangeon also recognized that involvement in World War I sharpened the Dominions sense of self-interest. The discussions in the New Zealand Parliament on the Discharged Soldier Settlement Bills point to a more subtle reality: the coexistence of a number of competing discourses of Empire and nation that persisted as the 1920s unfolded, at a time when New Zealand's economic dependence on Britain remained virtually absolute.

Notes

1   Apple growing was vigorously promoted from the 1910s in terms of an arcadian paradise, see Roche, M. (2003), 'Wilderness to Orchard: The Export Apple Industry in Nelson', New Zealand 1908-1940, *Environment and History,* vol. 9, pp. 435-450.

2   The New Zealand Labour Party was not founded till 1916 for convenience I have used labour' as a shorthand describing the affiliation of several MP's elected prior to this time who subsequently became part of the New Zealand Labour Party.

3   The *New Zealand Farmer and Stock and Station Journal* contained a series of articles on Land Settlement and Immigration in 1922 e.g. G.L.P (1922) Land Settlement and Immigration, A Comprehensive Scheme Wanted, *New Zealand Farmer and Stock and Station Journal,* 1 June, pp. 837-838; G.L.P. (1992) Land Settlement and Immigration II, *New Zealand Farmer and Stock and Station Journal,* 1 July, pp. 870-871.

4   David Buick the MP for Palmerston died in the 1918 influenza epidemic. The Crown subsequently purchased 273 acres at £68 from his estate for the Cloverlea Soldier Settlement nearby to Palmerston North. This proved to be not the most successful of settlements largely on account of the small section sizes (LS 1/21/183 Cloverlea Settlement Archives New Zealand, Wellington).

5   Kopane Settlement was subdivided from a larger estate and in that sense was a continuation of a process of closer settlement initiated by the Liberal government in the 1890s. Although the price paid for the land attracted much attention in parliament the local history of the district tries to argue that it was sold very cheaply to the Crown (see Roche, 2002).

6   As part of the anti German sentiment during the war some items of food such as 'German Sausage' and 'German Biscuits' were renamed, and still remain today 'Belgium [Sausage]' and 'Belgium Biscuits'.

7   Kauri forests (*Agathis australis*) were widespread throughout Northland. The species extrudes resin. Fossilised lumps of this known as Kauri gum could be dug from swamps

and were used to make paint and varnish. Lower grade gum including that bled from living trees was used in the manufacture of linoleum.

8   The Tainui settlement reached in 1995 involved $NZ170mill compensation, the return of some previously confiscated land still in Crown ownership, and an apology.

9   LS1/21/183 Cloverlea Soldier Settlement Archives New Zealand, Wellington. Gould (1992, Appendix 2) discusses how army nurses were deemed ineligible for land settlement assistance until 1919 as well as providing cameos of some half dozen who took poultry and dairy farms.

References

Anon. (1926), 'New Zealand', *The Roundtable*, vol. 16, pp. 655-71.

Anon. (1975), 'Thomas Sidey', New Zealand Biographies National Library of New Zealand, Wellington, vol. 2, p. 172.

Arnold, A. (1981), *The Farthest Promised Land*, Victoria University Press with Price Milburn, Wellington.

Bassett, M. (1995), *Coates of Kaipara*, Auckland University Press, Auckland.

Bennett, J. (2001), 'Maori as Honorary Members of the White Tribe', *Journal of Imperial and Commonwealth History*, vol. 29, pp. 33-54.

Brooking, T. (1996), *Lands for the People*, Otago University Press, Dunedin.

Brooking, T. (1996a), 'Use it or Lose it, Unravelling the Land Debate Nineteenth Century New Zealand', *New Zealand Journal of History*, vol. 30, pp. 141-62.

Burton, O. (1935), *The Silent Division: New Zealanders at War, 1914-1919*, Angus and Robertson, Sydney.

Christopher, A. J. (1988), *The British Empire at its Zenith*, Croom Helm, London.

Clout, H. (2004), 'Albert Demangeon, 1872-1940: Pioneer of La Géographie Humanine', *Scottish Geographical Journal*, vol. 119, pp. 1-24.

Condliffe, J. (1930), *New Zealand in the Making*, Allen and Unwin, London.

Constantine, S. (ed.) (1990), *Emigrants and Empire; British Settlement in the Dominions Between the Wars*, Manchester University Press, Manchester.

Demangeon, A, (1925), *The British Empire: A Study in Colonial Geography*, Harrap, London. translated by Ernest F. Row.

Duder, C.J. (1993), 'Men of the Officer Class': The Participants in the 1919 Soldier Settlement in Kenya', *African Affairs*, vol. 92, pp. 69-87.

Dunne, B. (1993) 'The Ideal and The Real. Soldiers, Families and Farming: 1915-1930', BA (Hons) Long Essay, University of Otago, Dunedin.

Fairburn, M. (1975), 'The Rural Myth and the New Urban Frontier', *New Zealand Journal of History*, vol. 9, pp. 3-21.

Fedorowich, K. (1995), *Unfit for Heroes, Reconstruction and Soldier Settlement in the Empire Between the Wars*, Manchester University Press, Manchester.

G.L.P. (1992), Land Settlement and Immigration, A Comprehensive Scheme Wanted, *New Zealand Farmer and Stock and Station Journal*, 1 June, pp. 837-8.

G.L.P. (1992), Land Settlement and Immigration II, *New Zealand Farmer and Stock and Station Journal*, 1 July, pp. 870-1.

Goodall, V. (1962), *'Flockhouse, A History of the New Zealand Sheepowners' Acknowledgement of Debt to British Seamen Fund*, Keeling and Mundy, Palmerston North.

Gould, A. (1992), 'Soldier Settlement in New Zealand after World War I: A reappraisal', in J. Smart and T. Wood (eds) *An ANZAC Muster: War and Society in Australia and New*

Zealand 1914-18 and 1939-45. Monash Publications in History No. 14. Monash University, Clayton Vic., pp. 114-29.

Gould, A. (1994), 'Proof of Gratitude? Soldier Land Settlement in New Zealand after World War One', unpublished PhD Thesis in History, Massey University, Palmerston North, New Zealand.

Gould, A. (2000), 'Soldier Settlements', in McGibbon (ed.) *Oxford Companion to New Zealand Military History,* Oxford University Press, Auckland, pp. 498-502.

Gould, A. (2000a), 'Massey, William Ferguson', in McGibbon, I. (ed.) *Oxford Companion to New Zealand Military History,* Oxford University Press, Auckland, p. 313.

Gould, A. (2000b), 'Maori and the First World War', in I. McGibbon (ed.) *Oxford Companion to New Zealand Military History,* Oxford University Press, Auckland, pp. 296-9.

Hamer, D. (1996), 'Hornsby, John. Thomas Marryat, 1857-1921'. *The Dictionary of New Zealand Biography, Vol. Three,* Auckland University Press, Department of Internal Affairs, p. 231.

King, J. (1998), 'Anti German Hysteria During World War I', in Bade, J.N. (ed.) *Out of the Shadow of War,* Oxford University Press, Melbourne, pp. 19-24.

Lester, A. (2002), 'British Settler Discourse and the Circuits of Empire', *History Workshop Journal,* vol. 54, pp. 25-48.

Leneman, L. (1989), 'Land Settlement in Scotland after World War I', *Agricultural History Review,* vol. 37, pp. 52-64.

Lockwood, C. (1998), 'From Soldier to Peasant? The Land Settlement Scheme in East Sussex 1919-1939', *Albion,* vol. 30, pp. 439-62.

LS 1/21/183 Cloverlea Soldier Settlement, Archives New Zealand, Wellington.

Meaney, N. (2003) 'Britishness and Australia: Some Reflections', *Journal of Imperial and Commonwealth History,* vol. 31, pp. 121-35.

Miller, C. and Roche, M. (2003), 'New Zealand's 'New Order': Town Planning and Soldier Settlement after the First World War', *War and Society,* vol. 21, pp. 63-81.

Montgomerie, D. (2003) 'Twentieth Century New Zealand War History at Century's turn', *New Zealand Journal of History,* vol. 37, pp. 62-79.

Morrell, W.P. (1935), *New Zealand,* Benn, London.

New Zealand Parliamentary Debates, Government Printer, Wellington (various years).

Ogborn, M. (2000), 'Historical Geographies of Globalisation, c.500-1800', in B. Graham and C. Nash (eds) *Modern Historical Geographies,* Pearson Education, Harlow, pp. 43-69.

Olssen, E (1996), 'Paul, John Thomas, 1874-1964', *The Dictionary of New Zealand Biography,* vol. Three, Auckland University Press, Department of Internal Affairs, pp. 390-92.

Parr, J. (1926), 'The High Commissioner for New Zealand, 1926', *United Empire,* vol. 17, pp. 601-604.

Pawson, E. (2004), 'The Memorial oaks of North Otago: A Commemorative Landscape', in G. Kearsley, and B. Fitzharris (eds), *Glimpses of a Gian World, Essays in Honour of Peter Holland,* School of Social Sciences, University of Otago, Dunedin, pp. 115-31.

Phillips, J. (1989), 'War and National Identity', in D. Novitz, and B. Willmot, (eds), *Culture and Identity in New Zealand,* GP Books, Wellington, pp. 91-109.

Phillips J. (1990) '75 Years after Gallipoli', in Green, D. (ed.), *Towards 1990: Seven Leading Historians Examine Significant Aspects of New Zealand History,* Government Printer, Wellington, pp. 91-106.

Phillips, J. (2000), 'National Identity and War', in McGibbon, I. (ed.) *Oxford Companion to New Zealand Military History,* Oxford University Press, Auckland, pp. 347-50.

Powell, J.M. (1971), 'Soldier Settlement in New Zealand 1915-1923', *Australian Geographical Studies*, vol. 9, pp. 144-60.

Powell, J.M. (1980), 'The Debt of Honour: Soldier Settlement in the Dominions, 1915-1940', *Journal of Australian Studies*, vol. 5, pp. 64-87.

Pugsley, C. (1995), *Te Hokowhitu A Tu, The Maori Pioneer Battalion in the First World War*, Reed, Auckland.

Roche, M. (2002), 'Soldier Settlement in New Zealand after World War I: Two Case Studies', *New Zealand Geographer*, vol. 58, pp. 23-32.

Roche, M. (2003), 'Wilderness to Orchard: The Export Apple Industry in Nelson', New Zealand 1908-1940, *Environment and History*, vol. 9, pp. 435-50.

Sharp, M. (1981), 'Anzac Day in New Zealand 1916-1939', *New Zealand Journal of History*, vol. 15, pp. 97-114.

Sinclair, K. (1986), *A Destiny Apart: New Zealand's Search for National Identity*, Allen and Unwin, in association with the Port Nicholson Press, Wellington.

Taylor, A.J.P. (1972), *Beaverbrooke*, Hamilton, London.

Trlin, A.D. (1979), *Now Respected Once Despised, Yugoslavs in New Zealand*, Dunmore Press, Palmerston North.

Worthy, S. (2004), 'Light and Shade: The New Zealand Written Remembrance of the Great War, 1915-1939', *War & Society*, vol. 22, pp. 19-40.

Chapter 9

'Oriental Sore' or 'Public Nuisance': The Regulation of Prostitution in Colonial India, 1805-1889

M. Satish Kumar

In a play by Vijay Tendulkar (1996) entitled, *Gashiram Kotwal*, based on Marathi history and conventions, there is a depiction of the life of Balaji Janardan Bhanu (1742-1800) who became an administrator or Phandnavis at the age of 14 years by hereditary right. In this play the life of Ghasiram, a Brahman turned Kotwal (or Chief Police Officer of Pune) is interspersed with a vivid description of Poona city. The play highlights the dominance of the Police Raj (empire) known for its endemic brutality and corruption. The Police Raj served as a metaphor for the colonial regime (Arnold, 1986, p. 2). In other words the police became a mainstay of colonial power. This police power was used to subvert and circumvent legal process in order to administer prompt retribution and collective punishment. The colonial police became the corner stone for the institutional realisation of the Benthamite ideal of the Panoptican in urban and rural India. The Police Raj became known for its brutal methods of repression. Being subordinate to the civil administration and responsible for ordering rural and urban society, it was largely immune from public prosecution for its excesses (Arnold, 1986). The following is a dialogue that ensued in the city of Poona one morning in 1778.

Gashiram Kotwal declares that 'to kill a pig, to do an abortion, to be a pimp, to commit a misdemeanour, to steal, to live with one's divorced wife, to remarry if one's husband is alive, to hide one's caste, to use counterfeit coins, to commit suicide, *without a Permit* is a sin. A good woman may not prostitute herself; and a Brahman may not sin, *without Permit'*. Permits were an invention of the Police Raj and were a precursor to all forms of Regulation introduced in the British Empire. Here, prostitutes too were issued a permit or a ticket after they were incarcerated and examined in the Lock Hospitals, scattered in various geographic locations particularly in the military spaces of the Empire. Thus, given the state of medicine in the 19th century, arguably venereal disease was 'more a matter of police than of medicine' (Ballahatchet, 1980, p. 30).

Prostitution emerged as one of the biggest threats to the civilizing, moral agenda of the British Empire. The location of the British Empire in the tropics presented a fertile space for the emergence of discourses and tropes linking sexuality, race and identity to the ever-increasing imperial drive for superiority and dominance. In this context, prostitutes were referred to as a threat, as 'outcastes' and an aberration from the 'normal'. They were held responsible for spreading venereal disease, the *'feringhee'* sore that had become a public nuisance and a source of embarrassment for the colonial government. This was also a threat to the newly emerging middle class bourgeois sensibilities within the major metropolises in the new colonial order. Prostitutes were seen as dirty, disorderly and transgressive. They represented an embodied profession (based on their body) and an embodied practice (based on the establishment of a patron-client exchange). This chapter argues that prostitutes exhibited a contested, discursive identity in the colonial mentality, which led to their spatial bounded-ness and confinement. But despite being victims of tradition, patriarchy and colonial order, prostitutes were able to carve out spaces for resistance based on their agency. They took advantage of the ambivalence inherent in the colonial order's support for state-regulated sex to the military and used it to their advantage. Prostitutes indeed, can only be understood in a specific cultural context and it is therefore important to appreciate the various forms of cultural representations associated with prostitution in the colonial context. Such cultural representations were inevitably undercut by debates between the need for the domestication of 'fallen' women and for the naturalization of their identity based on their social, and functional utility. Colonial ambivalence therefore stemmed from the State's inability to choose between naturalization and domestication.

This understanding reinforces the need to move beyond the simplistic approach of branding all prostitutes in homogenous terms. Prostitutes constituted a complex category within the British Empire in India, which did not lend itself easily to explanations, and most of the studies alluding to prostitution generally tend to provide an Occidental understanding of the subject. There is a need to appreciate and understand the incredible variety of identities subsumed under the generic term 'prostitute'. This variation stemmed largely from the individual's class and religious or caste affiliation, and from their rural or urban context. Consequently, prostitution as understood within a largely liberal western, post-colonial construct has a completely different cultural connotation to the Indian context. Any attempt at a standardized definition of what a prostitute is or prostitution meant in the colonial context introduces cultural and spatial biases and prejudices, which may not adequately reflect the specific context in which prostitution was and is practised.

This chapter aims to reconstruct the spaces and regulatory measures of prostitution and thereby chart its historical geography in colonial India. It also analyses the discourses of morality and racial prestige in these prostitutional spaces. It seeks to show that regulation was not the recent invention of British colonialists but that it had existed as an accepted part of the cultural ethos since ancient times in India. By locating prostitution in this way, we decentre the binary presumption that the colonies were always at the receiving end of metropolitan 'home' injunctions. Indeed the greater the adaptation of existing cultural mores and norms by colonialists the more this helped to establish a fluid space of interaction, where actors and institutions interacted at a reciprocal level in the colonial context (Levine, 1996; Cooper and Stoler, 1997; Howell, 2000; Thomas, 1994). However, most colonial records present some problems in tracing these interactions, insofar as they tend to be largely reflective of official prejudices and the socio-cultural values of the prejudiced. For example, in June 1887 an official note prepared by the Government of India and reported in the Sanitary files, observed that '... the cantonment authorities can keep... the immediate vicinity of barracks free from the outcastes, who are so dangerous to the men'.[1] A further report from 1893 stated, 'with the single exception of the beef-butchers, prostitutes are the class of all classes whose residence it is most important to regulate strictly. Regard for public decency requires that they should not be allowed to pursue their vocation in all parts of the cantonments indifferently, more especially as owing to the nature of their calling, they would naturally select (as they often do in cities) the most public thoroughfare for its practice... Accordingly, we find that in all cantonments, public prostitutes are required to reside in certain specified portions of the bazaars; and at Lucknow a European harlot was not long ago turned out of the bazaar and left the cantonment, because she declined to comply with this rule'.[2] We can notice from this text the explicit endorsement of spatiality and spatial bounded-ness associated with prostitution and the effort made to justify this by presenting a moral case which was righteous and impartial.

Prostitution as a discursive domain had a marginal place in the cultural topography of the colonial empire. Considering the ways in which such discourses have been construed both in the empire and the colony, it is clear that the indigenous discourses ran parallel to the discourses which emerged in the colonial period. Here prostitution was viewed differently by indigenous and colonial civil society. Prostitution organized in the colonial urban space was closely reflective of the gendered notions of colonialism, where the liberty to intervene in the social space of the colonial civil society was largely constrained by the politics of race and class. More than this, it also subverted the binary opposition of privileged masculine, and feminine identities, and the dis-privileged sites of control. Modernity no doubt, produced a dichotomized discourse of the female subject as a 'good or bad woman'. At the same time, colonialism while demystifying this subject also subsumed this discourse in the larger body of miasmatic disease and dirt, thereby creating its own racialized sexual politics of gendered control. The period 1860 to 1890 saw the politicization of this colonial subject 'prostitute'.

Colonial discourses were closely intertwined with medical discourse. While colonies came to be looked upon as a vast pathological reservoir, prostitutes came to be perceived as representative of this malaise. The enlightened colonial discourse on prostitution swung between coercion and persuasion, representing it as a site of submission, resistance and interaction between the natives and colonizers. As a site of interaction it provided a context for complete dominance. At the same time while coming into conflict with issues of morality and class, it generated rumours, suspicions and even hostility. Here indigenous discourses of the 'Oriental' body and prostitution were grafted onto the colonial hegemonic discourses to suit a particular project of civilization and control. This chapter foregrounds how discourse and identity were intertwined in the colonial, cultural construction of the prostitute. This site of performance for the prostitutes was also the site of resistance from the hegemonic discourses of the 'Other', of race and class.

Prostitution was also looked upon as a necessary social evil. This idea when transplanted into the colonies had a differential impact on indigenous civil society. Prostitutes were viewed as victims of poverty, and at the same time represented the private failings of their moral character (Levine, 1994). In fact, they were looked upon as 'sinners'. In this sense, prostitution reflected gender, class and racial biases. Thus, the regulation of colonial urban space in South Asia was contingent on such discourses, whereby issues of prostitution was conveniently conjured as being symptomatic of the decadent culture prevalent in these societies. It became a matter of State policy to intervene in the sexual relations between a prostitute and the client. Domestication of the prostitutes was an important social and moral agenda of the state. As a result, poor, urban working-class women were particularly targeted. Public health came to be equated with public order. Within this realm of South Asian *habitus* the negotiation between the structural imperatives of the colonial power and subjectivities related to morality, ideology, and power was taking place. Responsible for generating cultural perceptions and the shared social dispositions that inform actions in everyday space, the notion of *habitus* is central to any understanding of the regulation of prostitution in the context of social relations in South Asia. In this context both the colonialist and the indigenous male population sealed their indictment of the prostitutes based on shared patriarchal values.

The system of regulation through which they did so is discussed at length below. We simply note here that this allowed colonial authorities and the middle class representatives of the native population to draw on the discursive resource of an ideology of domesticity to give themselves social legitimacy, rights and privileges over any women suspected of being a prostitute. Despite the fact that the spatial boundaries for these prostitutes were circumscribed by specific orders, they managed to negotiate from within the interstices of these specific hegemonic representations. Like other pre-colonial socio-economic formations, the profession of prostitution underwent a dramatic change in India from a pre-colonial to a colonial identity. It attracted women from a variety of socio-economic strata in society and also attracted a new clientele from the *'nouveau rich'* who were products of the colonial order. At the same time, prostitution was subjected to control and surveillance and was affected by the new commercial economic relations and new ethical social norms introduced by colonialism throughout the 18th and 19th centuries.

Authorial Representation of Prostitutes

The terms used by prostitutes to describe themselves are indicative of their class and caste background. Those of the higher caste would refer to themselves as 'patita' or 'fallen' women, referring to their more genteel past. Those from the lower order of the society referred to themselves as sex workers or *'tawaif'*, *'naikeens'*, *'thakahi'*, *'beshya'*, *'kanki'*, *'randi'* or *'ranrh'*, the emphasis being on the commercial, transactive identity (Banerjee, 2000, p. 8; Gupta, 2001). This is an important distinction, which is not reflected in some recent works on colonial sexuality (Levine, 1994; Ogborn, 1993; Peers, 1998; Mahood, 1990, Hyam, 1990; Howell, 2000; Stoler, 1996). The point is can we provide a historical and social agency for the prostitutes in colonial India? *Tawaifs* or prostitutes were socially acknowledged for maintaining moral order and purifying towns/cities as outlets for men's sexual drive.

Sin versus Crime

The prostitute's body became a site for contestation and controversy in the colonial period. Prostitutes became a subject of national debate in India's colonial cities by virtue of graduating from their traditional image of social sin in pre-colonial times to one of criminality under the colonial administration. The caste-based, hierarchical pre-colonial India accommodated the 'sinning' prostitutes as it did other socially functional occupational groups, such as sweepers, without identifying them as a 'criminal'. However in the colonial period prostitution was branded as a crime to be codified regulated and controlled. Its codification as a crime was shaped by orientalist perceptions, whereby the colonial concept of 'crime' clashed with the indigenous concept that allowed 'sinners' some space in society, albeit grudgingly. This clash of perceptions was made explicit in the observations of a British police officer posted to India during the last decades of the 19th century. Acknowledging the prevailing indigenous societal norms he said:

> Prostitution in India must be viewed from a different standpoint to that which we are accustomed. Prostitutes are treated with a degree of respect, are tolerated and even encouraged in India, to an extent incomprehensible to western standards of ethical thought... (Somerville, 1929, p. 1).

While damning prostitution as 'vice and crime', the British officer however ignored the fact that the taverns, brothels, opium dens and gambling saloons were fall-outs of the colonial order of modernity, and gradually all of these occupations came under the colonial surveillance and regulation. As part of this, obscene acts and songs were banned by the Indian Penal Code of 1860, and the clause was used indiscriminately against prostitutes and their trade. However, by 1888, the House of Commons in its resolution of the 5 June of that year, noted that 'prostitution is regarded by society in India in a very different light from that in which it is looked upon in England. We do not say that it is even in India a reputable or a honourable calling. But the disrepute, which attaches to it, is wholly different, in kind as well as in degree, from that which attaches to it in Europe'.[3]

An Historical Geography of Prostitution in India: Identity and Hierarchy

As Howell notes, 'Regulation existed in Britain well before the introduction of the Contagious Diseases Act of 1864. This was also true of India. In the 4th century BC, Kautilya in his Arthashastra laid down norms and rules as to how prostitutes should deal with their customers. It also provided for punitive measures like fines to be imposed on customers who cheated on prostitutes or harmed them in any way. The tax paid by the prostitute was important revenue for the state. Prostitution was seen neither as a crime nor a sin, but was treated like any other trade' (Howell, 2000, p. 324). Its practitioners were expected to abide by the regulations formulated by the administration, which also protected them from acts of misdemeanour and violence. Thus between 321-296 BCE, Kautilya's Arthashastra laid particular stress on regulation. According to him a Revenue Collector would 'Spy on the arrival and departure of men and women of condemnable (anarthya) character, as well as (on the) movement of foreign spies' (Shamasastry, 1929, p. 159).

In the ancient texts and mythologies of India the role of the prostitutes was clearly defined in society and was not circumscribed by injunctions stemming from a moral high ground. The position of the prostitutes was systematically legitimized during war when they fell victim to rape and abuse. During natural calamities, when women were abandoned by families, they took to the vocation as prostitutes and as concubines in royal households or as *devadasis* (temple dancers) or as mistresses in a Brahmin household. The prostitutes participated in religious and social functions. As Banerjee (2000, p. 24) notes, 'prostitutes had a special position during sacred religious ceremonies. In fact, clay collected from the earthen floor of a prostitute's threshold (*beshyadwara-mrittika*) was used in religious ceremonies. The logic being that this soil was the purest as it contained all the accumulated virtues, which were shed by the men who lost them once they entered a prostitute's room'. This is adequately immortalized in Bollywood films such as Devdas, where 'Chandramukhi', the prostitute is asked to provide the soil for the annual religious ceremony.

Through the medieval period from the 13th to the 17th centuries, prostitutes were provided with similar protection and status. The life styles of these courtesans were dramatized and eulogized in various forms of narratives in popular culture. Prostitutes, referred to as courtesans or *tawaifs* (be it the temple dancer or *devadasis,* or concubine such as Anarkali or Umrao Jan in Moghul India, or Chandramukhi in Hindu royal households) received state patronage and protection for their service to the public. The functions of courtesans were well defined and they remained oppressed and sexually exploited within the stranglehold of a traditional patriarchal system. Unlike the public women or commercial prostitutes these courtesans were captives within the feudal households. They were to cater to the sexual needs of the patriarchal society. As ancient practitioners of prostitution, courtesans had numerous skills like music, dancing, and painting and had the space to cultivate these skills with a great deal of freedom. This was unlike the prostitutes who emerged in the new colonial order. Thus, we find a gradual colonial secularization of the identity of a prostitute from that of the earlier religious ideal. Later developments in Hinduism saw lawmakers turning prostitutes into criminals while at the same time patronizing the profession as a social necessity. We find a parallel in that, just as in the ancient India, so too in the 19th century, British colonialists who came with the objective of cleansing native society ended by patronizing and supervising prostitutes. A finely graded moral scale informed all acts of regulation in the colonial period.

Oldenburg (1984) noted that there were class and religion-based hierarchies among the traditional prostitutes, particularly among those drawn from the Hindu and Moslem communities. The religious divide in the later part of the 19th century became accentuated when Hindus viewed Muslim as dirty, unclean and therefore disorderly (Gupta, 2001, p. 112). Thus *'tawaifs* or *baijis'* were distinct from *'khemtawalis'* (Banerjee, 2000). Likewise, race became a significant marker too in distinguishing European and Eurasian prostitutes set apart from indigenous prostitutes in the newly emerged colonial order. The classification of prostitutes into various hierarchies was done to protect middle class bourgeois (*bhadralok*) sensibilities and at the same time, to maintain a veneer of respectability and avoid social discomfort.

The composition of prostitutes underwent constant change over time. Prostitutes who dominated the red light market in the late 18th and early 19th centuries were largely from wide catchments of displaced women widows and abandoned wives and daughters from middle class households. They were replaced in the second half of the 19th century by hereditary prostitutes and by women coming from chronically poverty stricken areas, or belonging to depressed castes who remained victims of natural calamity such as famines, floods and earthquakes. Thus lower class prostitutes replaced the earlier domination by those of the upper classes. Banerjee (2000) lists at least five categories of prostitutes in 1860. Out of 6871 registered prostitutes in 1872, Hindus numbered 5804 and Muslims 930, while the rest were English, Irish, Russians, Austrians, Poles, Hungarians, Italians, French and Spaniards. Prostitutes were also segregated by their rural or urban place of origin, thereby catering for diverse levels of clientele. The number of prostitutes multiplied greatly during natural calamities and this was reflected in the imperial censuses. Women in impoverished households resorted to prostitution as a temporary measure to supplement family income. The colonial moral order did not allow for social support to be accorded to female victims of natural and familial calamities. Thus colonialism fractured both public and private space. During this period, celebrated Victorian notions of 'gallantry and adultery' were superimposed within both the military and civilian spaces of prostitution. This was an accepted rhetoric visible among the military, traders, administrators, planters, and indeed adventurers of all shades. The brothel or 'chakla' became by default a symbol of respectability for the ambivalent colonialist perched on the moral high ground as well as for the newly emerging indigenous bourgeois, desperate to seek and mimic the bygone social grace and status of the traditional feudal landlords of pre-colonial India. Thus complicity in the practice of prostitution remained unchallenged despite opposition by moralist and purity campaigners. Acquiring a mistress was considered prestigious and honourable despite the later enforcement of 'Victorian morality', which led to rigid norms of spatial segregation between the colonizers and native population. This period also saw the emergence of a commercially institutionalized prostitution.

A History of the Regulation Acts

Colonialism gradually led to the institutionalization of a hierarchical system in both the civilian and military spaces of prostitution. The Regulation Acts designed to control the trade embodied a continuing tension between enshrining the conventions pursued in pre-colonial India, accommodating the new colonial moral order and adjusting to the constant resistance and conflict associated with this forced imposition. Therefore prostitutional spaces were targeted to bring the whole trade under colonial surveillance and control. As mentioned, Permit Raj, in operation since early 18th century, was the precursor to the Regulation Acts. In a singular sense, 'governmentality replaced territoriality'. Specific legislation was enacted to cover both active and dormant prostitutes in the form of Cantonment Act of 1864, amended in 1880, for military spaces, and the Contagious Diseases Act of 1868 for civilian spaces. This became essential because of the failure to control the rise of venereal disease among the troops and the need to enforce greater surveillance and control of prostitutes in military spaces.

There is evidence of continuity in regulatory practices from pre-colonial India into the colonial period. In 1780s, Samuel Hickson, a soldier in the East India Company noted that surgeons in Bombay were authorized to regularly inspect the public women of the bazaars and forcibly detain them for treatment if they were found to be infected with venereal disease.[4] The procedure of mandatory inspection and treatment of diseased prostitutes was formalized in 1805 in all military cantonments in India in order to check the spread of venereal disease. This method became a precursor to the establishment of Lock Hospitals and was seen as being more effective since it did not lend itself to the later charge of coercion (Peers, 1998, p. 150). Subsequently it was readapted as the Contagious Diseases Act in 1868. Between 1805 and the 1860s, Lock Hospitals became the mainstay in cantonments for checking venereal disease. They were established in 1807 in Bengal and Madras on the recommendation of Lord Bentinck, the Governor of Madras (who was himself suspected to have been infected; Peers, 1998: 150). It is interesting to note the emergence of Lock Hospitals at the same time in Glasgow (1805), Newcastle (1813), Manchester (1819), Liverpool (1834) and Leeds (1842; See Peers, 1998). Perhaps the idea of the Lock Hospital came from the Empire (India) to Great Britain in the early 19th century (Fischer-Tine, 2003, p. 177). This is speculative but is suggested by indirect evidence showing that East India Company officials, while surveying the existing judicial procedures in India in the 18th century, came to the conclusion that prevailing Indian practice of levying fines for sexual misdemeanours was purely extortionate and reflected judicial venality rather than a legitimate exercise of 'lordship'. Thus the judicial reforms of 1772 abolished the existing *faujdari bazi jumma* or fines in existence during the Islamic rule of India. The Company felt that complete withdrawal of punitive sanctions against immorality might defame its administration in the eyes of the natives, since British rule might then be taken to have encouraged immorality.[5] Arguably, this may be the reason why lessons learnt from India may have helped to endorse Victorian morality in England and the rest of the Empire. The initial decision to establish Lock Hospitals did not result in a hue and cry, and the greatest support came from the military. However with the mounting pressure from purity campaigners and abolitionists in London and the Empire against the practice of state sponsored and regulated sexuality (and supported by Bentinck), the Lock hospitals were suspended in 1830. Bentinck believed in the importance of reiterating the moral and medical agenda rather than focusing on the issue of surveillance, but a lack of finance due to the Anglo-Burmese Wars of 1824-36, and the rising graph of venereal disease were also cited as reasons for their abolition.

Regimental officers of the East India Company army led the campaign for the establishment of Lock Hospitals (Peers, 1998, p. 150). As early as 1806, medical police were in place, unofficially, to oversee the health of prostitutes and the soldiers. It was unanimously accepted that soldiers could not be policed. Accordingly, the prostitutes were obliged to bear the brunt of regulation in the regimental bazaars and cantonments in order to control venereal disease. As we have seen, one way to do this was to attach all prostitutes who consorted with the soldiers to various regiments and cantonments.[6] Thus the efficacy of Lock Hospitals was established by the fusion of medical and military discourses on the protection of European soldiers. The 1805 resolution decreed the regulation of prostitutes by establishing Lock Hospitals in all cantonments which served as a base for European soldiers. Monthly identification and inspection of prostitutes was resorted to and attempts were made to ensure that the Lock Hospitals did not encroach upon the territory of sacred spaces, such as temples. Provisions were made for infected women in these hospitals.[7] Evidence suggests that 17 Lock Hospitals were in operation in Madras as early as 1808 with 3,502 women in attendance.[8] Freedom to cohabit with native women resulted in a rise in the incidence of venereal disease in 1815. The only way forward was to expel diseased prostitutes rather than accept that the regulation policy had failed. Those prostitutes who refused to enter the Lock Hospital were tonsured and drummed out in public (Raj, 1993). These expulsions were designed to send a strong signal to those who were not willing to accept their disease and be treated.[9] This was an idea borrowed from continental French practice of 1635. By the 1820s most Lock Hospitals were placed under the supervision of the senior surgeon in the cantonments, with the hospital matron being given charge of their day-to-day running.

Peers has shown that after the abolition of the Lock Hospital system in 1830, clandestine measures continued to be adopted by the military to control venereal disease in the cantonments. This took the form of informal free check ups for prostitutes (Peers, 1998, pp. 628-47). However, the lack of effective medical control and the lack of legal options to oust diseased prostitutes led to an impasse.[10] Consequently, despite the ban, Lock Hospitals were gradually re-established on an informal basis in the cantonments and camps in Madras and Bombay in the 1840s and 1850s. This time punitive action was taken against diseased prostitutes and severe fines were imposed for those afflicted with venereal disease. Thereafter, London gradually relented and allowed the formal reopening of the Lock Hospitals on a trial basis for a year, before the Contagious Diseases Act of 1868 was formally introduced in India and rest of the Empire. Metropolitan agendas were spatial in character and were constantly reinvented and justified on the basis of experiments conducted in the colonies, thereby reinforcing the division between the civil and military spaces of interaction. Venereal disease was abstracted from a medical condition to inform social and moral discourses in the colonies. Thus in India, the regulation of sexuality became the focus of intense debate from 1805 until the abolition of Contagious Diseases Acts in 1892. Once again gender as a category was collapsed on to racial attributes, contrasting the '*manly*' British military and the '*effeminate*' native Indians. With the Contagious Diseases Act of 1868, a formal distinction was made between vice management and control of disease in the Empire. Thus surveillance and regulation, commonplace in the metropolitan centres in Europe, was already in operation in the colonies before the formal introduction of the Contagious Diseases Acts.

The Cantonment Act of 1864

In 1864 the Imperial Government promulgated Act XXII, known as the Cantonment Act for enforcing the 'inspection and control of public houses dealing with prostitution and for preventing venereal disease'. The Act effectively divided prostitutes into two classes, those who were and were not frequented by European soldiers. The former were designated as *first class* prostitutes and were subjected to the regulations envisaged by the Act. Here protection of the British soldiers was of paramount importance. Army authorities set up chaklas in regimental bazaars and other appointed places in the cantonment. Soldiers found to be infected were fined five rupees and this was duly deducted from their salary.

The Imperial Indian Contagious Diseases Act (Choudda-Ain), of 1868

Following the 1864 Contagious Diseases Act in England, in 1868 the *Choudda-Ain* or Contagious Diseases Act was implemented in India, with the aim of registering prostitutes and providing for their examination. It was introduced in Madras in October 1868, and in Bombay in May 1870. According to Stewart and Peile, the Act gave the government power to regulate for the inspection and control of brothels and for the prevention of the spread of venereal disease.

> The rules so made directed that a register of public prostitutes should be prepared and kept in the office of the cantonment magistrate; that a woman wishing to practice prostitution was required to apply to have her name entered on the register (Table 9.1); that no woman not so registered was allowed to practice prostitution; that on registration the woman had to assent to certain regulations which were explained to her; that every registered prostitute had to pay a monthly subscription to the Lock Hospital Fund; that she was furnished with a printed ticket to be renewed annually that she was bound to present herself with her ticket for medical examination at the Lock Hospital not less than once a fortnight, that if found diseased she was detained in the Lock Hospital until discharged as cured (Figures 9.1 and 9.2). That if she wished to change her residence in the cantonment she was bound to notify it; and that she was to observe rules made by the cantonment to maintain cleanliness in houses occupied by registered prostitutes and any prostitute convicted of a breach of these rules was liable to a fine of 50 rupees or imprisonment with or without hard labour for 8 days.[11]

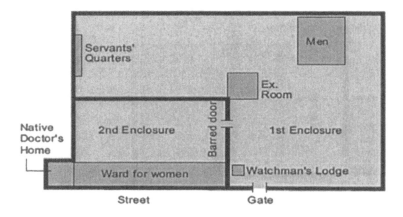

Figure 9.1 Lucknow Lock Hospital

The Act initially came as a relief for the harassed prostitutes in the cantonments. They came forward willingly to register and undergo medical inspection in order to pursue their profession without hindrance.[12] At Lucknow and Meean Meer, each girl paid nine pice to a pleader for writing out the application required before entry to the chakla. For the first time a distinction was made between regimental and commercial or common prostitutes. The Act initiated compulsory registration of prostitutes and subjected them to periodic examination and treatment in all the cantonments usually on a biweekly basis. The movement of the prostitutes was regulated in specified spaces. If a soldier became infected he was questioned as to the identity of the woman and she was hunted up and sent to the Lock Hospital. All registered prostitutes were provided with a ticket, which included their name, caste and residence. They were obliged to present their ticket when challenged by the police; failure to do so risked imprisonment.

Table 9.1 1892: Ticket of Registered Prostitutes in the Cantonment of Meean Meer

Name	Begum 1st
Caste	Mahomedan
Registered number	1
Place of residence in Cantonment	Sudder Bazaar
Date of Registry	4.3.90
Personal Appearance	

C.L.M.Rich
Officiating Cantonment Magistrate,
Meean Meer

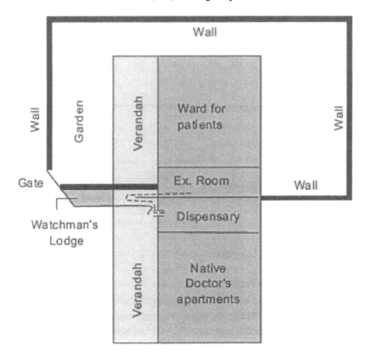

Figure 9.2 Sitapur Lock Hospital 18 April 1893

The Contagious Diseases Act very quickly became embroiled in controversy because it proved incapable of reducing the incidence of venereal disease and led to great excesses and harassment. One issue was its inability to distinguish between hereditary prostitutes, mistresses or even those who would occasionally prostitute themselves, especially during natural calamities. A further bone of contention concerned the finance for the Contagious Diseases establishments. Sharing of expenses was a major headache. Thus the moral high ground as far as the control of the vice of prostitution was concerned, was balanced by the problems inherent in the Act and its harsh implementation.

The 'Body' therefore became the site of control, resistance and complicity within the colonial rhetoric of hygiene, civilization and culture. Like all other forms of colonial codification and classification, the body too became a context for codification along with the sexual mores of the period. This was the codification of the client–prostitute relationship, which was often couched in explicit, gendered notions. Thus for men it was an expression of the 'irregular indulgence of a natural impulse' whereas for the prostitute it was an offence, a sin committed for gain (Walkowitz, 1980, pp. 69-71). The Contagious Diseases Act ensured the segregation and confinement of prostitutes into specified spaces of cities and towns and any attempt at violating these regulations meant heavy penalties.

Although the Contagious Diseases Act repeated the principal features of the existing Cantonment System, it went far beyond the confines of the cantonments. Its remit covered prostitutes from rich and poor backgrounds and targeted courtesans and hereditary prostitutes. Surveillance was the only mechanism to help prevent venereal disease from taking on an endemic form in the cantonments. The Contagious Diseases Act was initiated to control the activities of commercial prostitutes and prevent their access to soldiers. The Cantonment Act sought to maintain native prostitutes exclusively for the British soldiers and they were kept in protective custody as captives in the cantonment chaklas. In this sense regimental prostitutes were considered to be of a special class. It was observed that native harlots lost their caste among their peers when they consorted with British soldiers, as the latter were regarded as men of low caste. Accordingly, *'gora kamana'*, to earn one's living from the British soldier was used as a term of reproach.[13] Natives were prohibited and practically prevented from consorting with women in chaklas which were frequented by British soldiers.[14] It is interesting to note that a class system was in operation among the European as well as among the indigenous prostitutes. Bombay boasted a three-tier system in 1912 with first, second and third class brothels.[15] This concept was clearly adopted from the old regimental system in operation in the military cantonments and was imported into the civilian sphere.

The 1893 Report by General Sir D.M. Stewart and Sir J.B. Peile, observed that prostitutes accompanied the regiments on marches and in the standing camps.[16] Four years earlier, in 1889, it had also been noted that regiments coming to the cantonment 'brought their women with them'. The main intention of this system was to ensure that all women consorting with British soldiers were 'fit and healthy'. In 1890, the Lancer's chakla, the Derbyshire Regiment chakla, the West Kent, Gordon Highlanders, West Yorkshire and Argyll and Sutherland Highlanders chakla, and the Royal Irish chakla, were all marching from Peshawar to Ambala. These chaklas were regulated spaces in the empire and were attached to specific regiments. The prostitutes were graded and classified, and payment for their services was determined by their physical beauty, race, class and religion.

Chaklas

In the chaklas the prostitutes were accommodated under one roof, sometimes in one bazaar or central place and sometimes in separate chaklas in different parts of the cantonment.[17] The chaklas were large irregular buildings about 125 feet long with very high walls, and windows high up in the wall, barred with wood. The prostitutes paid two rupees a month in rent for each room.[18] Contemporary sketches by Dr. Kate Bushnell of the Lucknow and Peshawar chaklas (Figures 9.3 and 9.4), indicate 20 rooms, arranged in suites of ten on each side. In these chaklas no rent was paid by the prostitutes as they belonged to the government.[19] Prostitutes like any other profession or trade were listed as an occupation in the register of the cantonment. Chaklas were gloomy places of confinement and ill treatment. Most chaklas were repulsive looking, the rooms occupied by the women being merely flat roofed, mud-built apartments, containing little more than a pallet.

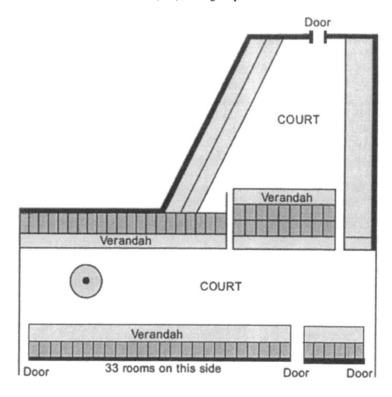

Figure 9.3 Lucknow Chakla

The rooms in the Sadr Bazaar chakla were numbered in English, denoting the prostitute's registration number and the general cantonment number of the house. Originally this was intended to enable a soldier to identify a woman otherwise than by her name, and insofar as these rooms were still occupied by prostitutes these numbers continued to serve this purpose.[20] The chaklas resorted to by British soldiers were inspected by the military police who patrolled the bazaar. They were instructed to remove any native men from these quarters, a fact which implied extraordinary vigilance with women being under guard at all times. It was also noted that 'if a woman failed to present herself for examination the medical officer reported the fact to the cantonment magistrate, who probably fined her for non-attendance and ordered her to attend or leave the cantonment'. Besides a 'woman who had left the cantonment could not practise prostitution again until she had been examined'.[21] There was also evidence that soldiers subjected most of the prostitutes to physical abuse and ill treatment when they were drunk.

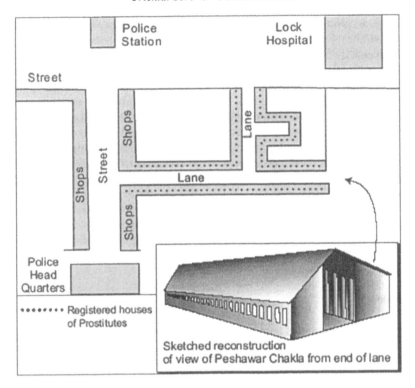

Figure 9.4 Peshawar Chakla

Payment for services rendered by the prostitute was largely governed by the military rank of her client and her own class as a prostitute. A uniform minimum rate of pay applied in all chaklas, A corporal paid eight annas, a bombardier six annas, and a private four annas. If they refused to pay, the money was deducted from their salary.[22] The provision of girls for the chaklas also involved payment. One of the '*Mahaldarni*' (matrons) stated that she used to be paid five, ten or 20 rupees to get girls for the regiment. 'For a very attractive girl she was furnished with 50 rupees'.[23]

Agency and Resistance

Gender and sexuality were thus used as the basis for establishing cultural and moral hierarchies in which Europeans were ranked above the Indians (Peers, 1998, p. 149). Nonetheless, the prostitutes retained an immense capacity for agency and constantly subverted the colonial designs of regulation and control of their activities. They took on various guises as religious mendicants, singers, and milkmaids. Some of them went on to pretend to be married and had their names taken off the registers. A racial divide was clearly established between indigenous and European prostitutes and their treatment in the Lock Hospitals reflected this. Some prostitutes were keen to get themselves arrested, concluding that they would be imprisoned for only a month, whereas if they were found infected during their weekly check ups, they would be incarcerated for six months leading to loss of income. Despite the police repression and constant harassment from cantonment authorities, prostitutes continued to use space strategically to evade arrest and conviction. They operated from outside the spatial zone where the regulation was in force, tempting soldiers to consort with them outside the confines of the chaklas. Many managed to leave the Lock Hospital before the completion of their treatment. Although local newspapers were divided in their opinions in most presidencies of India, yet there was a common agreement that the Contagious Diseases Act was detrimental to the modesty and wellbeing of the prostitutes in particular and women in general (Ramana 2002, p. 164). Prostitutes in the Empire managed to undermine patriarchal sensibilities by deploying space and culture to their advantage, thereby destabilizing and hybridizing the private and public moral/immoral spaces of interaction.

Conclusion

Prostitution as an embodied subject was highly context dependent. The perception and sensibilities associated with prostitution varied across class, race and space. Prostitutes were not a homogenous category and authorial representations showed them to be a highly varied community. The introduction of a colonial moral order tended to further dichotomise the subject 'prostitutes'. Their traditional identity as a 'sinner', accommodated within the social spaces of pre-colonial society, was over turned, and they were branded as criminals, outcasts from society. This ran contrary to the indigenous perception and created great ruptures.

The historical geography of prostitution clearly demonstrates the hierarchy that prevailed in the pre-colonial order, which represented a totally different moral scale from that introduced by British colonialism. At the same time regulation was not a colonial invention. It existed as far back as the 4th century BC. However, a process of secularization of prostitute identities did occur during the colonial era, which transformed these from their previous religious ideal. This resulted in a class system emerging within the prostitutional order in the 18th and 19th century, reinforcing what was essentially a Victorian moral order.

The Regulatory Acts demonstrate how governmentality replaced territoriality and how 'space' became instrumental in ordering, regulating and sanitizing the prostitutes' world. A distinction needs to be made between the Cantonment Act or 'Old-System', which had its origins in 1805, and the Contagious Diseases Act of 1868. Before the latter was abandoned in 1888, it had succeeded in institutionalizing and commodifying an entirely new hierarchy of prostitutes, and these were significant in raising the level of discourse to new plane. In effect, a new class of professional regimental prostitutes emerged, who had the status of aristocracy within the occupational group. They had more pay, but fell in esteem among their less favoured compatriots. Thus the Contagious Diseases Act tried to control and regulate public women, whereas the Cantonment Act tried to groom a new class of prostitutes. There were innumerable instances of agency being invoked by prostitutes who were resisting control and regulation. They constantly took advantage of the ambivalence of the colonial moral order and manoeuvred successfully from one crisis to another. Prostitutes in the colonies were in a captive system and did not have a freedom that their counterparts had in Britain. Thus the chaklas were prison-like and under constant surveillance. Each prostitute was coded, numbered and classified, and to their soldier clients might be as anonymous as the rooms they occupied. Medically inspected and licensed to practice, each prostitute acquired an identity as the property of a given regiment, In this dehumanizing process, we see the ambivalent responses to issues of gender and sexual identity inherent in the construction of a colonial military *mentalité* of sexuality. An imperially inspired moral scale was structured and imposed on indigenous sensibility, and prostitutional spaces were organized and regulated in the new colonial urban spaces to cater for these new colonial sensibilities and perceptions.

Acknowledgements

The author acknowledges comments from the editors of the volume and to the prolonged discussion with Phil Howell in Cambridge during my stint as Commonwealth Fellow at the University of Cambridge. I also benefited immensely from the discussions with Gerry Kearns and Ulf Strohmayer in the European Social Science History Conference, March 2004, Berlin. Thanks are also due to Maura Pringle for cartographic assistance in the School of Geography, Queen's University, Belfast.

Notes

1   British Military Department (1887), 8 October 1886, File Nos 34-58, Home Sanitary, June, Oriental and Indian Office Collections (hereafter OIOC), Madras Archives.
2   Home Department (1893), Proceedings of the Government of India, Sanitary, Nos 7-9, pp. 9, 235, OIOC, Cambridge University.
3   Ibbetson, D., Cleghorn, J. and Khan, M.S., 1893, *Report of the Special Commission appointed to inquire into the Working of the cantonment regulations regarding Infections and Contagious Disorders, 2 June,* Resolution No. 2439-D, Military Department, p. 229, OIOC.
4   IOR (1781), MS Eur b 296/1, December, p. 79, OIOC, Madras Archives.
5   Buchanan-Hamilton (1814), Mss Eur D.91, IOR, p. 12. See also Smith (1915), OIOC, Madras Archives.
6   IOR (1808), F/4/2341, OIOC, Madras Archives. See also *Madras Medical Quarterly Journal*, 1839, vol. 1, p. 444.
7   Military Consultation (1805), vol. 335, 22 April, pp. 2267-2673, OIOC, Madras Archives.
8   IOR (1808), F/4/345, OIOC, Madras Archives.

9 Military Consultation (1815), vol. 586, 11 September, p. 10287, OIOC, Madras Archives.

10 IOR (1831), F/4/1338 and 53031, OIOC, Madras Archives.

11 [Separate Report by] General Sir D.M. Stewart and Sir J.B. Peile (1893), *Report of the Committee, Appointed by The Secretary of State for India, to inquire into the Rules, Regulations and Practice in the Indian Cantonments and elsewhere in India with regard to Prostitution and to the Treatment of Venereal Disease* [min. 2656-2665], p. xxvii, OIOC, Nehru Memorial Museum and Library.

12 Government of India, Public Department (1872), Government Order, 13 December.

13 Government of India, Military Department (1893), Resolution No. 2439-D, 2 June, p. 230.

14 Russell, G.W.E. (1893), *Report of the Committee to inquire into the Rules, Regulations and Practice in the Indian Cantonments and Elsewhere in India, with Regard to Prostitution and to the Treatment of Venereal Disease* (London), Evidence to the Special Commission, lines 2723-28, OIOC, Cambridge University.

15 Oriental and India Office Collections, IOR (1913), Government of India, Judicial Department, Government Order 2520, 6 December (Confidential).

16 Government of India, Military Department (1893), Resolution No. 2439-D, 2 June, p. xxviii, OIOC, Madras Archives.

17 Russell (1893), *Report*, xxxviii: p. 6.

18 Departmental Committee (1893), *Appendix I: Minutes of Evidence taken before the Departmental Committee appointed by the Secretary of State for India, to Inquire into the Rules, Regulations and Practice in the Indian Cantonments and Elsewhere in India, with Regard to Prostitution and to the Treatment of Venereal Disease at the India Office, Whitehall,* 11 April, p. 6, OIOC.

19 *Ibid*, p. 17, Discussions with Dr Kate Bushnell and Mrs E.W. Andrew, OIOC, Cambridge University.

20 *Report of the Committee (1893), Appointed by The Secretary of State for India, to Inquire into The Rules, regulations and Practice in the Indian Cantonments and Elsewhere in India with Regard to Prostitution and to the Treatment of Venereal Disease,* Min 2656-2665, p. xi.

21 [Separate Report by] General Sir D.M. Stewart and Sir J.B. Peile (1893), Report of the Committee, Appointed by The Secretary of State for India, to inquire into the Rules, Regulations and Practice in the Indian Cantonments and elsewhere in India with regard to Prostitution and to the Treatment of Venereal Disease, M.R. 3156-3159, OIOC.

22 Departmental Committee (1893), Appendix I: Minutes of Evidence taken before the Departmental Committee appointed by the Secretary of State for India, to Inquire into the Rules, Regulations and Practice in the Indian Cantonments and Elsewhere in India, with Regard to Prostitution and to the Treatment of Venereal Disease at the India Office, Whitehall, 14 April, p. 17, OIOC, Cambridge University.

23 Colonel Plowden to the Report (1893) of the *Special Commission Appointed to Inquire into the Working of the Cantonment Regulations Regarding Infectious and Contagious Disorders,* Y24265, p. 15, OIOC, Cambridge University.

References

Arnold, D. (1986), *Police Power and Colonial Rule: Madras 1859-1947*, Oxford University Press, Delhi.

Ballhatchet, K. (1980), *Race, Sex and Class Under the Raj: Imperial Attitudes and Policies and their Critics, 1793-1905,* Weidenfeld and Nicolson, London.

Banerjee, S. (2000), *Dangerous Outcast: The Prostitute in 19th Century Bengal*, Seagull Books, Calcutta.

Cooper, F. and Stoler, A.L. (1997), 'Between Metropole and Colony: Rethinking a Research Agenda', in F. Cooper and A.L. Stoler (eds), *Tensions of Empire: Colonial Cultures in a Bourgeois World*, University of California Press, Berkeley, pp. 1-56.

Fischer-Tine, H. (2003), 'White Women degrading themselves to the lowest depths: European networks of prostitution and colonial anxieties in British India and Ceylon ca. 1880-1914', *The Indian Economic and Social History Review*, vol. 40, pp. 163-90.

Gupta, C. (2001), *Sexuality, Obscenity, Community: Women, Muslims and the Hindu Public in Colonial India*, Permanent Black, Delhi.

Howell, P. (2000), 'Prostitution and Racialised Sexuality: The Regulation of Prostitution in Britain and the British Empire before the Contagious Diseases Acts', *Environment and Planning D: Society and Space*, vol. 18, pp. 321-39.

Hyam, R. (1990), *Empire and Sexuality: The British Experience*, Manchester University Press, Manchester.

Levine, P. (1994), 'Venereal Disease, Prostitution and the Politics of Empire: the Case of British India', *Journal of the History of Sexuality*, vol. 4, pp. 579-602.

Levine, P. (1996), 'Rereading the 1890s Venereal Disease as Constitutional Crisis in Britain and British India', *The Journal of Asian Studies*, vol. 55, pp. 595-612.

Mahood, L. (1990), 'The Magdalene's Friend. Prostitution and Social Control in Glasgow, 1869-1890', *Women's Studies International Forum*, vol. 13, pp. 49-61.

Ogborn, M. (1993), 'Law and Discipline in Nineteenth century English State Formation: The Contagious Diseases Acts of 1864, 1866 and 1869', *Journal of Historical Sociology*, vol. 6, pp. 28-55.

Oldenburg, V.T. (1984), *The Making of Colonial Lucknow, 1856-77*, Princeton University Press, Princeton.

Peers, D. (1998), 'Soldiers, Surgeons and the Campaigns to Combat Sexually Transmitted Diseases in Colonial India, 1805-1860', *Medical History*, vol. 42, pp. 137-160.

Raj, M.S. (1993), *Prostitution in Madras: A Study in Historical Perspective*, Konark, Delhi.

Ramana, M. (2002), *Western Medicine and Public Health in Colonial Bombay, 1845-1895*, Orient Longman, Hyderabad.

Shamasastry, R. (1929), Kautilya's Arthasastra, Mysore, Wesleyan Press, 3rd edition.

Smith V.A. (1915), *Rambles & Recollections of an Indian Official by Major General Sir W.H. Sleeman, KCB.*, Humphrey Milford, London. (Revised Edition.) (First published 1844).

Somerville, A. (1929), *Crime and Religious Belief in India*, Thacker Spink and Company, Calcutta.

Stoler, A. (1996), *Race and the Education of Desire: Foucault's History of Sexuality and the Colonial Order of Things*, Duke University Press, Durham.

Tendulkar, V. (1996), *Gashiram Kotwal*. Translated from Marathi by J. Karve and E. Zelliot, Seagull Books, Calcutta.

Thomas, N. (1994), *Colonialism's Culture: Anthropology, Travel and Government*, Princeton University Press, Princeton.

Walkowitz, J.R. (1980), *Prostitution and Victorian Society*. Cambridge University Press, Cambridge.

Chapter 10

Prostitution and the Place of Empire: Regulation and Repeal in Hong Kong and the British Imperial Network

Philip Howell

Introduction

Recent research into the historical geography of British imperialism and colonialism might be summed up in the twin questions: *when* and *where* was Empire? This is not being deliberately obtuse. The basic definitions of British (initially *English*) imperialism, quite apart from its significance and ultimate value, remain indistinct, and perhaps more so now than ever.[1] Problems of language and terminology are clearly a major issue here (Firth, 1918; Adams, 1922; Canny, 1998; Howe, 2000), but so too are difficulties in identifying the substantive temporality and spatiality of the Empire. The *time* of empire is clearly one problem: the origins of British imperialism, colonialism and colonization (policies, processes and practices that are often conflated) remain murky, whilst the apparently much more straightforward demise of empire has been challenged by critics of the British (post-)colonial present (Gikandi, 1996; Jacobs, 1996; Baucom, 1999). But the *geography* of British imperialism, once so conveniently mappable, has also become indistinct. It is not just the longstanding debate on Britain's 'informal empire' that is at issue here. The notion of a singular Empire is also difficult to reconcile with the ever-growing appreciation of the complexity and diversity of imperial rule and colonial practice: by the 1890s, as Peter Burroughs (1999, p. 171) notes, 'rather than constituting one empire, this conglomeration of large land masses and territorial fragments comprised several empires – pluralistic and singular in their individual dealings with Britain'. But perhaps most crucially of all, the once self-evident boundary between Britain and its Empire, the domestic metropolis and the colonial periphery, has been subjected to a series of critiques that have demonstrated the simultaneity and co-constitution of nation and empire. Thus, as Philippa Levine (2003, p. 4) succinctly puts it, 'Empire and metropole were not separate sites; empire was not a single site'. It has now become something of a commonplace to argue that the historiographical divide between metropolis and empire is artificial and misleading (Hall, 1994; Grewal, 1996; Cooper and Stoler, 1997; Sinha, 2002).

This chapter reflects these recent orientations in the study of British imperialism and reflects on the displacement of empire that has resulted. What is offered here is a consideration of one particular but notable colonial project – the regulation of prostitution or sex work – taking the British colony of Hong Kong as a key case in point. Prostitution regulation has rightly come to be recognized as a critical prop of imperial rule, vital not only to the campaign against venereal diseases but in a wider sense to the maintenance of racial and sexual privileges upon which colonial authority depended (McLintock, 1995; Stoler, 1996; Bryder, 1998; Stoler, 2002; Levine, 2003). This chapter recognizes and reinforces these arguments. What I aim to do here is to add to the literature on the relationship of sexual regulation to imperialism, not just in British Hong Kong but also in many other modern colonial regimes. But I also ask: what is the *place* of empire in this history of sexual regulation? The locality is of course one answer: I discuss in some detail the nature of British 'regulationism' in later 19th-century Hong Kong, along with its local geography – the landscape of regulated sexuality, its debt to discourses of race and cultural difference, and its embedding in the material practices of British colonialism in South East Asia. But I also want to link these local geographies to the spatiality of the wider British 'imperial network' (Lester, 2001), and to consider the ways in which sites like Hong Kong were constituted in and through their relationship with each other and with the mother country. British regulationism was much more than an archipelago of regulationist regimes; there really *was* an imperial system, a medical and political paradigm that informed practices throughout the Empire. In this chapter, I argue that it was British humanitarian 'repealers', vehemently opposed to regulationist ordinances, who made the greatest headway in both revealing and constructing this imperial system; they were the ones who most fervently insisted on linking colonial sites to each other and to the British metropolis. Ultimately we can say that their view of the place of empire triumphed, sealing the demise of regulationist practices not just in Hong Kong but across the British imperial world.

Using Hong Kong therefore as a springboard rather than simply as a case study, I want to consider not so much *colonial* forms of regulationism differing markedly from domestic practices and experiences but a practice that was truly *imperial*, instituted more or less simultaneously in a series of colonial *and* metropolitan spaces – that is, in the imperial network. I outline here the discursive and material geographies in which sex work regulation, and its humanitarian opposition, were implicated, emphasizing the links between colonial and domestic sites, but also going beyond these to the international legal and political regimes in which even the British Empire was but one part.

Prostitution and the Place of Empire

By regulation and 'regulationism' are meant the policies that evolved in 19th-century Europe to combat the epidemic threat of venereal diseases, syphilis in particular. The classic forms of regulationism, emerging in France from the Napoleonic period, sought to contain the ravages of disease by isolating and managing the professional sex workers who were considered to be the major conduits of infection within society (Harsin, 1985; Corbin, 1990; Aisenberg, 2001). By enforcing a strict system of state and municipal registration of prostitutes, regularly inspecting them for signs of disease, and locking them up in special lock wards or hospitals if they were found to be in a contagious state, European states hoped to minimize the dangers of sexually transmitted diseases. Of course, such policies reflected class, gender and sexual norms, with the result that working-class women were considered at the same time more culpable and more tractable to medical discipline than their male clients. In response to these inequities, there arose a counter-offensive from feminist and other critics of regulation – most notably in Britain, which had during the era of the Contagious Diseases Acts (1864-69) a diluted but no less significant regulationist regime (Walkowitz, 1980; Taithe, 1992; Ogborn, 1993). These 'repealers' took the fight not only to continental regimes, but also to colonial forms of regulationism, for sex work was also regulated in many British and European colonies (Ware, 1969; Taithe, 1992).

The British colonial experience of regulationism is particularly notable, and has emerged very clearly in a wide range of studies (Ballhatchet, 1980; Van Heyningen, 1984; Warren, 1993; Levine, 1994; Whitehead, 1995; Manderson, 1997; Banerjee, 1998). What is particularly significant about the British case is that although domestic regulationism was at best half-hearted (by French and continental standards at least) *colonial* regulationism was extensive, thoroughgoing and highly resistant to political pressure. Such forms of regulation stand out even amongst the markedly illiberal and authoritarian projects of Crown Colony government. They form an instructive contrast to domestic policy. What is more, British colonial regulation of prostitution predated the Contagious Diseases Acts by some, sometimes many, years (Ware, 1969; Howell, 2000). They were certainly not exported to the colonies from the metropolis. This does not necessarily mean that colonial sites should be identified as laboratories of disciplinary modernity, but it does suggest that the colonies played a much greater role in British policy than has been hitherto realised. Figure 10.1 shows some of the sites in which British regulationism was introduced: contagious diseases legislation is truly an empire-wide phenomenon.

Now no one has recently done more to elucidate these issues than Philippa Levine. In her comparative research on Hong Kong, the Straits Settlements, Queensland and British India, Levine (2003) is able to demonstrate that prostitution and its regulation were central to colonial administration and rule. She details the ways in which these colonial sites implemented legislation in accordance with their local economic and political situations, and, in particular, with their distinctive racial and sexual regimes. But these colonies also shared characteristic features, concerned as they all were with the management of sex. The 'attention to spatial detail' (Levine, 2003, p. 297) in colonial legislation is especially striking: the enclosure of sex work in the brothel and in brothel districts, the simultaneous privacy and public visibility of managed prostitution, the demand for isolation and containment tempered by the equally insistent desire to counter the unhygienic and impenetrable spatiality of the native quarters. And above all, there is the role of racial and sexual boundaries in shaping a distinctive geography of segregation. This geography differs markedly from the domestic situation, where brothels were never licensed and made central to the management of the sex trade, and where racial distinctions were never invoked with such systematic attention. Ultimately, therefore, Levine paints a convincing portrayal of British colonial regulationism as driven by anxieties about race, sexuality and imperial security.

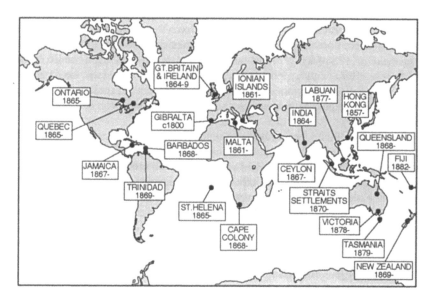

Figure 10.1 Regulationist Legislation in Britain and the British Empire[2]

In so doing, however, Levine largely reinforces the analytical separation between British and colonial policies. Hers is certainly the first full-scale trans-national study of prostitution regulation, and no one has covered such a range of colonial sites in such depth and detail. But it is not in fact empire-wide.[3] In each of Levine's case studies it is the sexual relations between 'white' settlers and sojourners and racialised 'others' which posed the critical problems regulationist policies were intended to address; but not all of the Empire was characterized by these problems, and thus not all forms of British colonial regulationism. In the 'near' Empire for instance – the Mediterranean colonies of Malta, the Ionian Islands and Gibraltar – the nature of contagious diseases legislation is quite different. In these colonies, small but significant for imperial security, regulationism was built on ethnic exclusions, national identities and sovereign claims, rather than on the racial categories adopted in the far-flung territories.[4] And in the nearest colony of all – that of Ireland – prostitution regulation clearly takes on a very different shape and significance; racial discourse is not exactly absent here, but it is clearly not comparable to what we see in the tropical colonies

(Wills, 1996; Luddy, 1997; Howell, 2003).[5] Thus the place of empire in the British project of sexual regulation is actually a far more difficult issue than it first appears; it cannot be reduced to the 'colonial' – no more than it can to particular colonial sites. If we consider the British Empire as a whole, we are likely I think to come to rather different conclusions about the management of sexuality.

We do not yet have such a history of British regulationism. What I would suggest about such a history, however, is two-fold. Firstly, we have to follow to the letter Ann Laura Stoler's (1996) dictum that metropolis and colonies should be held under a single conceptual aegis. Rather than invoking *the* colonial – as a singular, if nevertheless locally inflected – project, discourse and experience, we should actively trace the workings of an imperial culture, one that as a matter of course includes the metropolitan territory. And I would suggest here the utility of the concept of *imperium* – a formulation that might serve to remind us of the need to avoid essentializing 'the colonial' and identifying this space with particular colonial sites. Secondly, though it follows directly from this observation, instead of following the methods of comparative history it would be more useful to take up Alan Lester's (2000, 2002, 2002a) invitation to consider the operation of 'imperial networks'. That is, to consider the movement of people, commodities, projects and practices, and of ideas, information, knowledges and discourses from one colonial site to another and from each colonial site to the metropole (Hall, 2002; Lambert and Howell, 2003). Only by following through the promise of this approach will I think we be able to properly locate the Empire.

Now this is an ambitious agenda, and one still in its infancy; but I hope in the rest of this chapter to indicate some of the directions this approach can take, using the example of prostitution regulation in Hong Kong to flesh out these abstractions.

Regulationism and Colonial Governmentality in Hong Kong

Hong Kong has the distinction of having the earliest formal imperial legislation for the regulation of prostitution in the British Empire (Miners, 1987, pp. 170-190; MacPherson, 1995, 1997; Levine, 1998, 2003). Within 20 years of its occupation, a system of licensed prostitution was agreed between the local regime and the Colonial Office. The first piece of legislation, Ordinance 12 of 1857, required brothel owners in the colony to register their houses with the Registrar General, and made it compulsory for prostitutes to attend at the Lock Hospital if medical examination showed that they were in an infectious condition. The immediate impetus was the threat of venereal disease to the armed forces stationed in a colony described by Rear Admiral Stirling as nothing less than a 'Pest House' threatening British forces not only in the locality but also in the whole of the East Asian theatre (MacPherson, 1997, p. 88). In the military mind, British soldiers and sailors were the unwitting victims of the thousands of sex workers who were drawn to Hong Kong from South East Asia and the mainland of China, attendant upon the migrant male workers whose coolie labour was essential to the colony's prospects. By 1865, the European population of roughly 4000 was dominated by the Chinese population of over 80,000. Under these conditions, apprehensions about the extent of venereal disease in the services were hard to separate from wider fears of crime, disorder and the wholesale breakdown of imperial authority. The venereal disease ordinance of 1857 must be set amongst a series of measures introduced by Governor Bowring to ensure the safety and security of British rule. Christopher Munn (2001) has well documented the demise of the early, optimistic vision of 'Anglo-China', and its replacement by a much bleaker assessment of racial antagonism and mistrust. The machinery of police and bureaucratic surveillance that was created in these years amounted to nothing less than a 'reign of terror' (Munn, 2001, p. 280) for the Chinese population of Hong Kong, and the introduction of regulated prostitution must be understood with an eye to this crisis of government.

Certainly, the regulation of prostitution in Hong Kong was far more openly coercive and thoroughgoing than anything that could be contemplated in Britain at the time. In Hong Kong brothels were numbered and licensed, their inmates registered prospectively, rather than subsequent to arrest. The system of medical examinations relied upon brothel residence and brothel discipline for its efficacy. Special police were used to patrol unregistered houses and streetwalking women, their wages, along with the costs of a system of informers and the medical administration itself being met by a levy on brothel houses. These characteristics sharply distinguish colonial from domestic British practice; colonial discipline is in this sense a far more intrusive genus of disciplinary power. Most importantly of all, sex workers' clienteles were strictly segregated: brothels for Chinese men were distinguished from those for non-Chinese (predominantly European) clients. Race and racial privilege were thus built into the project of regulated prostitution in Hong Kong right from the start. This was a form of 'racialized sexuality' (JanMohamed, 1992) comparable with British and continental European practice elsewhere. Figure 10.2 gives an indication of the scale of this system; there were well over 100 recognized brothels in the late 1850s and early 1860s.

Figure 10.2 also indicates the change in policy that occurred under the administration of the reforming governor Sir Richard MacDonnell. Governor Bowring's venereal disease ordinance was repealed and replaced by the even more severe Contagious Diseases Ordinance 10 of 1867. This new piece of legislation placed the system of licensing firmly in the hands of the Registrar General, and allowed the police greater powers in prosecuting unregistered brothels and prostitutes. The renewed vigour that was introduced into the system of regulated prostitution is plain from the figures, which also show for the first time the numbers of brothels designated – as either 'Chinese' and 'foreign' – that is, for Chinese and non-Chinese clients.

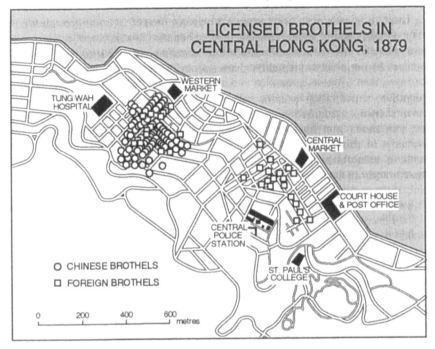

Figure 10.2 Recognised Brothels in Hong Kong, 1853-1878

The separation of the brothel clienteles was central to the functioning of this system, in both its 1857 and 1867 incarnations. It is entirely consistent of course with the discursive positioning of the non-European population as collectively, rather than individually, deviant and pathological. Within racial discourse it made perfect practical and philosophical sense to regulate the entire field of commercial sexuality by requiring registration, but confining medical examinations to those women who serviced European men. In the first, pragmatic, direction, the sexual health of the Chinese was simply not a priority for the colonial authorities. As one Registrar General was later to concede, 'We were not trying to save the Chinese from syphilis' (Hong Kong, 1879, p. 5). It is important to record, however, the fact that pragmatism concerning venereal disease extended to the concerns expressed by the Chinese community itself. The Chinese – or at least their representatives – deplored the extension of intimate medical examination to Chinese women, and though they could not prevent medical checks on women who slept with 'foreigners', they actively campaigned for other Chinese women to be spared this indignity. Their resistance to medical and bureaucratic surveillance helped circumscribe the impact of this form of colonial discipline. From the inception of Hong Kong's regulation of prostitution, these women working in Chinese brothels were allowed to practise their trade without British medical interference.

These pragmatic arguments against wholesale intervention into the circuits of commercial sexuality were, however, matched by the cultural relativist stances we have already noted. Colonial discourse accepted the Chinese community's aversion to the practice of intimate medical examination, and although this was portrayed as a Chinese 'prejudice' – a marker once again of irrationality and un-modernity – it meshed happily with the practicalities of colonial governance. Colonial policy here invested in a portrayal of the Chinese as a racial other whose very otherness represented a local reality to be recognized and engaged with. This discourse of difference is at the heart of arguments that put forward Chinese women's abhorrence of the genital examination as the central fact upon which regulationism in the colony was founded:

> It was a wise recognition of the natural laws at work in this mass of corruption called prostitution, when the Government of Hongkong confined the application of the principal provisions of the Contagious Diseases Ordinance, viz., compulsory medical examination, to the licensed prostitutes in houses for foreigners only, and exempted from the same law the great mass of prostitutes for Chinese (Hong Kong, 1879, p. 6).

These were not then apologetics; rather, they were part and parcel of a colonial expertise about Chinese culture that cautioned against hasty, inappropriate, and overambitious projects. Confidently proclaiming the virtues of local knowledge and expert wisdom, this kind of work placed certain kinds of restraints on the colonial project itself.

Cultural difference was thus recognised by colonial government. As such it represents something akin to the 'facts' that we can see restraining the interventionist programmes of 19th century European governments. Under the rubric of 'governmentality' Foucault and his later commentators have played up the role of such quasi-objects as 'society', 'economy' and 'population' in the disciplining of political ambition; but in the colonial field, the discourse of cultural difference – of 'race', to use the contemporary shorthand – might be considered in a very similar way. Colonial governmentality (Scott, 1995; Kalpagan, 2000, 2002), for all its authoritarian character, could still be marked by clear boundaries and limits.[6]

Chinese and 'foreign' clients were directed, after 1867, to two separate areas: the former to the western district of the city of Victoria, and the latter to the Central district. It is clear from the pattern of prosecutions of unlicensed brothels that the greatest government effort was expended on clearing the Central district of unauthorized houses, there being little point in policing unlicensed Chinese houses given their proscription of non-Chinese clients. But it has not been recognized that this geography of racialized sexuality was a product of the later, 1867 Contagious Diseases Ordinance. In the older Ordinance 12, brothels were more or less confined to the Chinese quarters of town (MacPherson, 1997, p. 90) – that is, to the western half of the city, and to the district of Tai Ping Shan in particular.[7] Ordinance 10 of 1867 *relaxed* this initial restriction by allowing brothels to be licensed in the formerly proscribed eastern districts. Since the residents of 'foreign' brothels were of course predominantly Chinese women, this amounted to a move away from the policy of confining undesirable Chinese to the western districts of Victoria. Cecil Clementi Smith, Registrar General of Hong Kong immediately prior to the introduction of Ordinance 10, represented the views embodied in the earlier ordinance when he remarked that 'As a rule I would only license houses in a street where Chinese only are living', adding that 'As to the location of brothels, the general principle is to keep them out of sight, where they shall not offend the respectable portion of the community' (Hong Kong, 1879, p. 2). That this view was comprehensively abandoned in the 1867 legislation can be seen from the comments of H.E. Wodehouse, the succeeding Registrar General. Wodehouse noted that 'with a view to keeping down the number of unlicensed brothels it would also be desirable that the licensed brothels should be as centralized, as accessible and as well-known as possible' (Hong Kong, 1879, p. 8).

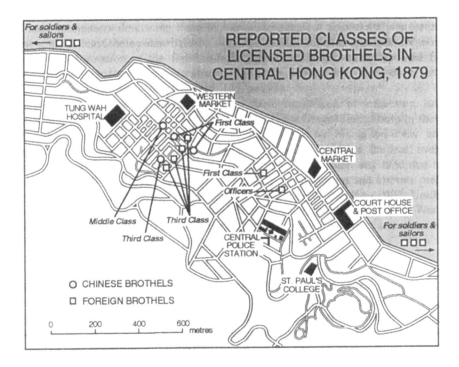

Figure 10.3 Registered Brothels in Central Hong Kong, 1879

The result was a much more completely regulated geography of regulated prostitution than had been possible in the 1857 legislation. And although race continued to play a crucial role in this system, more importance was attached to segregating brothel clienteles than to segregating Chinese and European society. There was too an explicitly acceptance here of class distinction, in ways that could cut against the ethos of racial segregation. Brothels both Chinese and 'foreign' were divided into first, middle, and lower-class houses, though it is notable that the 'best' Chinese houses on Hollywood Road were located only a few streets away from the counterpart third-class houses (Figure 10.4). By contrast, 'foreign' brothels were more segregated by their class of clientele. The lowest class of women was said to be to found in East Street and West Street in the far west of Victoria, in Tai Ping Shan, and in Wanchai. In Tai Ping Shan, in the face of the segregation of Chinese and 'foreign' brothels, there remained a number of licensed brothels for foreigners, catering to European soldiers and sailors and also to Malays. These small but thriving businesses were presumably relicts of the time when all brothels were confined to Tai Ping Shan and the western districts. However, the better class of women in the licensed brothels for Europeans were to be found primarily in the Central district (Hong Kong, 1879, p. 13). The owner of brothel No. 6 on Peel Street claimed for instance that 'my brothel is a first class one. No Europeans of inferior class go to my brothel; no soldiers or common sailors, and very rarely a Policeman': the brothel rather accommodated sea-faring men, captains and officers, and charged $2 a night, a rate comparable to the best Chinese institutions (Hong Kong, 1879, p. 28).

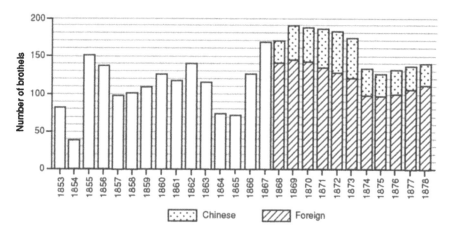

Figure 10.4 Reported Classes of Registered Brothels in Hong Kong, 1879

This landscape of commercial sex was undoubtedly produced by the characteristic preoccupation with race, and racial segregation is clearly visible. Yet attention must also be drawn to the shift in policy that occurred in 1867, a move away from the blunt abjection of Chinese from the European quarters that informed the earlier ordinance. After 1867 there was renewed attention to the licensing of brothels, but a different geography emerged. Chinese brothels continued to be confined to the western districts of the city, but 'foreign' brothels were allowed to encroach upon the Central district. Brothels for Europeans were spatially separated by class as well as by race, occupying a more diverse series of sites than had been possible under the earlier legislation. This was the result, on the ground, of the colonial discourse of race and sexuality, combined with the pragmatics of colonial rule. It represents an extension of disciplinary power, but one that recognized certain limits and opportunities. Regulationism in Hong Kong thus partook of a distinctively spatialized form of colonial governmentality.

Hong Kong in the Imperial Network

In Hong Kong, then, we see the workings of an exemplary regulationist project. From the earliest days of the colony disease prevention was of the highest importance, vital to the safety and security of British colonial rule. Prostitution, an inevitable accompaniment of Hong Kong's status as a military station and colonial entrepôt, was quickly identified as a major threat, and subjected to close administrative scrutiny as part of a package of measures designed to control the Chinese population. Venereal diseases were, however, never a purely medical matter. Disease prevention among the Chinese population was always a secondary issue, and the control of infected servicemen was likewise always secondary to the control of their Chinese partners. What is more, disease was identified with the Chinese population, their environments and their culture; management of the former depended upon the management of the latter. Racial discourse was fundamental to a system that embodied racial anxieties and entrenched racial as well as sexual privileges. This prophylactic project was characteristically strict and severe, yet we must also recognize its limits and its literal boundaries. Government of venereal disease did not, in this case, necessitate a wholesale surveillance and disciplinary control of the subject population – how could it, given the priority given to the colonizers' health and security? In Hong Kong, in what was regarded as a concession to Chinese 'prejudices' – though it accorded perfectly with colonial priorities – medical examination of prostitutes was limited to those with European clients, women who were first separated from their counterparts in brothels catering to Chinese men, and then segregated in a part of Victoria formerly reserved for European residents. Within the ambitions and inhibitions of colonial governmentality, just as the intention remained to manage venereal disease rather than to eradicate it, just so the Chinese population needed to be managed rather than comprehensively controlled.

In Hong Kong, this venereal paradigm, as we may call it, remained in place for half a century and more, its basic lineaments continuing to shape the colony's urban geography until well into the 20th century. Such a project testifies to the abiding imperial interest in everywhere regulating the sexual lives of colonial subjects. Regulation did not, in this case, necessarily mean intrusive intervention; it could also mean a calculated and diplomatic neglect, part and parcel of a 'wise' recognition of both cultural difference and governmental priority.

Yet the nature of regulationism in Hong Kong can be appreciated neither as a case study nor within the framework of a comparative history that is just as surely founded upon the proliferation of cases. To take this approach would in some senses be to replicate imperial circumscription itself. For the British, the avoidance of suggestions of system and centralism was almost second nature: in the imperial arena, localism was as much of a watchword as in domestic policy. Vulnerable in any case to charges that regulationism was a French or continental system illegitimately imported into Britain and to its possessions abroad, the British government put the stress on local necessity and the relative autonomy of colonial authorities to decide for themselves what measures were appropriate for their circumstances. Imperial lawmakers, instead of prescribing legislation, settled for encouraging and enabling colonies to pass their own local statutes and ordinances, including those for the control and management of prostitution. In these ways, the British government could insist that such practices did not amount to 'regulationism' at all. There was not, at least in the terms of plausible denial, any such thing as an imperial *system* of sexual regulation. This view may be characterized as the *official* discourse on regulationism.

Regulationism in Hong Kong could be justified not only because it was a response to local, but also to specifically *Chinese* conditions. It is well appreciated that prostitution in British colonies was effectively normalized, treated, in contrast to the home country, as an endemic feature of indigenous societies and their cultures. Hong Kong's experience exemplifies this wholesale 'Othering', in this case of the Chinese and their culture, allowing British authorities to claim both the necessity and the legitimacy of effectively licensing prostitution. In a passage to which I have drawn attention before (Howell, 2000, p. 332), the 1877 commission of enquiry into prostitution in Hong Kong argued that 'the principal points of difference between the various classes of Chinese prostitutes of Hong Kong and the prostitutes of Europe amount therefore to this, that Chinese prostitution is essentially a bargain for money and based on a national system of female slavery: whilst European prostitution is more of less a matter of passion, based on the national respect for the liberty of the subject' (Hong Kong, 1879, p. 5). For these commissioners, 'prostitution' in Hong Kong was *different*: in a culture where Chinese women were routinely bought and sold, where prostitution was more or less an accepted profession, where there was no basis for a moral or legal condemnation, prostitution was normal and had to be accepted as such. There is a clear echo of the statement made a few years earlier by the 1871 Royal Commission on the Contagious Diseases Acts, famous for its defence of the sexual double standard, that '[w]ith the one sex the offence is committed as a matter of gain; with the other it is an irregular indulgence of a natural impulse' (Parliamentary Papers, 1871, p. 17) – save for the fact that here both European prostitutes and their clients are described as driven by 'passion', whilst a whole culture and people is described as vicious and mercenary.

Now the 1877 commissioners were, in fact, largely hostile to the regulationist system in Hong Kong, not supportive of it. But there were supporters or prostitution regulation that took exactly the same line, arguing that domestic, liberal-minded objections carried no force in the colony's cultural and geographical situation. What is more significant though, and certainly less remarked upon, is the fact that in Hong Kong it was not just prostitution that was normalized; so too was the system of regulationism itself. In contrast to those who saw this, for good or ill, as an essentially European policy, there were important voices raised in Hong Kong to argue that regulationism was in fact a *Chinese* system, and one so ancient and ingrained as to be inseparable from Chinese culture itself. The 1877 commissioners carefully considered for instance a paper prepared by an officer of the Imperial Maritime Customs, on the 'Chinese Social Evil' (Hong Kong, 1879, pp. 54-60). This paper traced the development of state-organised prostitution in Chinese dynastic history, detailing the progressive acceptance of the necessity of regulating, rather than repressing prostitution, and emphasizing the contribution that this system made to the imperial revenue. Regulationism was indeed, therefore, 'imperial': but it was the celestial empire that originated it. If the British Empire adopted the system, it was only as a matter of colonial pragmatism, an unfortunate departure from essentially liberal British convictions, but one that was necessitated by 'Chinese' conditions. Not only was this a nod to what we now recognize as 'cultural relativism', it also stems from the specific crisis of government identified by Christopher Munn (2001).

That this was not an eccentric view is demonstrated by the fact that there was cross-partisan agreement about the essentially Chinese origins of prostitution regulation. Alfred Lister, Acting Registrar General under Sir Richard McConnell's administration, and a key proponent of regulationism, argued for instance that:

> It is impossible for the Government to look upon prostitutes as other than people who have deliberately chosen their mode of life, and generally mean to adhere to it. The peculiar social position of Chinese women, the way in which the Empire is overstocked with them, their uselessness, according to Chinese ideas, for any other purpose, our proximity to mainland, whence fresh supplies are always to be obtained, and the very small amount of shame which attaches to the calling, all make the work of diminishing the evil, otherwise than by regulating it, one which at any rate at present seems hopeless (Hong Kong, 1879, pp. 255-256).

And on the other side of the political fence, the 1877 commissioners more or less adopted this view wholesale, though distancing themselves from any suggestion of prostitution and its regulation as a necessary evil. Their report argued quite straightforwardly that prostitution regulation was not, in essence, a British innovation:

> Whatever vices the Chinese of Hongkong may have learned from foreigners, whatever evils the intercourse between foreigner and Chinese may have created, prostitution, brothels and the system of licensing brothels with a view to raise a revenue, legally or illegally, were indigenous institutions in China centuries before the present nations of Europe emerged from barbarism. When the Colony of Hongkong was first established, in 1842, it was forthwith invaded by brothel keepers and prostitutes from the adjoining districts of the mainland of China, who brought with them the national Chinese system of prostitution, and have ever since laboured to carry it into effect in all its details (Hong Kong, 1879, p. 3).

Inevitably, therefore, the nature of regulationism as a British policy becomes lost in the official view, overridden as it is here by the demands of an 'orientalist sociology' (Levine, 2000) and its cultural relativist priorities, as well as by those of colonial pragmatism and its invocation of local necessity. Regulation, therefore, whilst it might be 'imperial' in origin, is not here clearly recognised as a *British imperial* project.

That there was a system of British regulationism – an accumulation of knowledge, experience, precedent and practice that is imperial in nature and origin – is, however, quite clear. All over the British world, British rulers made characteristic recourse to the management of the Empire's sexual subjects (Ware, 1969; Taithe, 1992; Howell, 2000; Levine, 2003). Those who made the running in portraying regulationism as just such an imperial project were the *repealers* who had, since the late-1860s, put the British military and the British government firmly on the defensive. For these repealers – pushing, that is, for the repeal of the domestic Contagious Diseases Acts – the regulation of prostitution licensed vice, entrenched the despised double standard, and rode roughshod over individual liberties. They identified regulationism with continental tyrannies and moral laxity, and for all the British government's protestations, clearly regarded British practice as a domestic variant of essentially the same policy. Furthermore, they kept a watchful eye on the colonies, which they regarded as specially vulnerable, since British authorities could there introduce systems for the regulation of prostitution with little regard for either colonial subjects' liberties or for metropolitan opinion. As James Stansfeld (1882, p. 40) put it, the Hong Kong ordinances showed 'how far our rulers are prepared to go – when they dare'. They posed a danger not just to far away peoples with different coloured skins but to each and every British citizen and subject.

For repealers, the invocation of local necessity, the emphasis on the need to let the experts and the proconsular 'men on the spot' (Benyon, 1991) determine what measures should be employed in each colonial site, was absolutely rejected. For them, such a view was dangerous and exculpatory.

We might say that these very active men and women did not just reveal the system of imperial regulation practised by the British from at least the beginnings of the 19th century; in a sense, they helped, in their humanitarian work, to *construct* it. They achieved this in several ways. Activists like James Stansfeld (1882) were at pains, for example, to show that the British Colonial Office was wholly complicit in the 'government brothel system' in Hong Kong and elsewhere: there was no political distance between the metropolis and the periphery. In their comparative surveys (Deakin, 1872; Amos, 1877; Wilson, 1907), repealers also repeatedly drew attention to both imperial and international systems for the regulation of prostitution; their house-journal, the *Shield*, was particularly important in focussing attention not just on individual cases (see 24 May 1879 for their special issue on Hong Kong) but also on the system in which they were but elements. Repeal organizations also sponsored investigative tours, of which the Indian exposés of Elizabeth Andrew and Katharine Bushnell (1898) are the best known. In all this work, the repeal movement constantly represented local regulationist ordinances as part of a wider, corrupted imperial system. Not only was the movement for repeal an international campaign, therefore, an international 'humanitarian network', to borrow Lester's terms (2002); it also invested heavily in portraying British regulationism as the product of an imperial network. It focussed attention, for example, on Sir Henry Storks, ex-governor of Malta and a notable promoter of the highly regarded system of regulation there, and campaigned against him when he stood for Colchester in the by-election of 1870 (Hammond and Hammond, 1932, p. 151; Walkowitz, 1980, 106-7). As one of the subjected districts under the Contagious Diseases Acts, Colchester offered repealers a chance to campaign against domestic regulationism, but in making the link with the experiences of Malta, campaigners were able to present domestic parliamentary concerns as also necessarily connected to colonial, military ones. They tracked the movements not just of individuals like Storks, but also of policies, practices and prejudices. They put the lie to the official line that held the colonial sites in isolation from each other and from the metropolis.

These humanitarian opponents of regulation were thus both products and producers of the imperial network. Through their activities it was never possible in the public sphere to consistently portray British regulationism as a series of isolated, local ordinances, however much the British authorities would have preferred this. In this regard, for all their political marginality, repeal activists may always have had the upper hand. And when domestic repeal was finally achieved, in 1886, it is telling that Surgeon-Major Blair Brown could argue in these terms against the retention of contagious diseases ordinances of India: 'Surely the legislature of the English Parliament on such a subject is good enough for India? We hear of the unity of the Empire; in this there is little' (Brown, 1888, p. 80). These critics revealed an imperial network of regulated sexuality and demanded a unified moral and medical system. The repeal-minded were the ones who presented the most strongly unified vision of the empire. Repealers not only helped construct the idea and reality of an imperial network; in so doing, they were at the forefront of those who aspired to a humanitarianism that transcended the artificial frontier that separated the British metropolis from its colonies.

In contrast to what I have called official discourse, then, this humanitarian discourse endorsed the view that regulationist ordinances in colonial territories were more than just the sum of their parts. Regulationism could not, furthermore, be strictly divided into colonial and metropolitan variants. In contrast to the official view, which emphasized colonial difference, for the repealers regulationism was the same, imperial, legislation – the same outrage – wherever it took place and whomever it touched. Colonial and metropolitan forms of regulation were certainly distinctive, but they were not in principle separable.

To suggest that the repealers' view of the consistency of regulationist practices is closer to our own than to supporters of regulation is not, of course, to endorse their views uncritically. It is clear from Catherine Hall's recent work that we cannot accept the humanitarian network on its own terms. Hall (2002) is able to demonstrate how the oldest and most significant of these networks – that of anti-slavery – reconstructed as well as deconstructed the divide between metropolis and colony, remaining at its core racially inscribed and defensive of white privilege. But Hall's work – the most engaging demonstration so far of the possibilities of a new imperial history that takes the displaced geography of empire seriously – does offer a trans-imperial story of connection and complexity that corresponds to my argument here. For neither was regulationism exported from Britain to its colonies, nor imported to the metropolis from the colonial laboratories of modernity. Rather, it took its substance and significance from the movement of ideas, material, and people across the imperial network.

I have pointed above to the peregrinations of individuals like Sir Henry Storks, and the role that humanitarian networks played in revealing the significance of these imperial connections. But there is in Hong Kong a more significant example. David Lambert and I have suggested some of the ways in which the history of regulationism in Hong Kong can be related to the experiences of its ninth governor, Sir John Pope Hennessy, as he was caught up in both the imperial and humanitarian networks (Lambert and Howell, 2003). Pope Hennessy's antipathy to the system of regulated prostitution that was practised in Hong Kong can clearly be linked to his liberal, reformist humanitarianism, and to his 'pro-native' inclinations (Lowe and McLaughlin, 1992). Most importantly, Pope Hennessy's passionate anti-slavery convictions were confirmed during his immediately previous tenure as governor of Barbados and the Windward Islands, where he sided with the island's black population against a white plantocracy that he regarded as complicit in slavery-like practices: as he put it, in Barbados, 'the traditions of the slavery period are still powerful' (Lambert and Howell, 2003, p. 12). Coming to Hong Kong, therefore, Pope Hennessy was inevitably primed to understand the status of Chinese prostitutes as essentially that of slaves. This of course made it all the worse that the colonial government should take on the role of organizing the system; for him, government licensed brothels in Hong Kong were 'a means of keeping Chinese girls in a state of slavery' (Lambert and Howell, 2003, p. 15). Pope Hennessy took his usual decisive action, overriding local objections, for of course no British colony could tolerate the practice of slavery. Pope Hennessy was particularly exercized by the fact that one of the reasons that the system was introduced in the 1850s was to combat Chinese brothel slavery, and by the fact that, given that the government taxed brothels for Chinese clients, but did not examine their inmates, it was hard to avoid the conclusion that the government made money out of the trade. The British rulers of Hong Kong were little more than pimps and slave-masters.

All this is perfectly consistent with the repealers' views. Regulation could be read as complicit with the slavery that represented the antithesis of the benign model of empire. But regulationism in Hong Kong was also bound up with the status of the *mui tsai* ('little sisters'), young girls more or less sold into domestic service by poor parents (Miners, 1987, pp. 170-190; Jaschok and Miers, 1994; Pedersen, 2001). Notionally, this was a form of adoption or patronage, in which the *mui tsai* could be looked after, in return for service, until they came of age. But in practice this could lead to the meanest domestic drudgery and exploitation, or much worse. It could be linked to the buying and selling of female children for the purposes of prostitution. Sexual trafficking was another kind of outrage, and for some in Hong Kong, like Pope Hennessy's Chief Justice Sir John Smale, just as clearly a form of slavery. But Pope Hennessy found himself caught in a dilemma: his humanitarian sympathies, on the one hand, argued for intervention and elimination of this 'evil'; his sympathy for the subject population, on the other, argued for acceptance and non-interference. In the end, Pope Hennessy chose to follow the arguments of his protégé, the sinologist Ernest Eitel, to the effect that it was rash and illegitimate to label these practices as 'slavery' (Hong Kong, 1882; Lambert and Howell, 2003, pp. 15-19). Thus we find complex distinctions being made in Hong Kong, in which prostitution and its regulation was enmeshed; 'brothel slavery' could be condemned, whilst the related question of the *mui tsai* was only a 'so-called slavery'. The universalizing tropes of anti-slavery discourse and the humanitarian network were finding real barriers and boundaries in late 19th-century Hong Kong.

In linking these questions to the operation of imperial and humanitarian networks, David Lambert and I have attempted to portray slavery as a concept and a discourse that travelled and unravelled with the movements of imperial agents such as Pope Hennessy. With the specific question of prostitution regulation in mind, we can see how this travelling discourse, together with the networks in which it was embedded, conditioned the opposition to regulation at a local level. The very meaning of prostitution regulation was contested. It was a thoroughly unstable concept. Was it a French, British, or Chinese policy? Were contagious diseases ordinances severally or systematically related? Were colonial and metropolitan policies related, or quite distinct? Could it be labelled a form of slavery, or was it unwise to indulge in such condemnations? These were, and are, geographical questions. For what I have called the official view, a geography of isolated colonial sites, empowered to determine their own needs and to derive their own contagious diseases legislation, was authorized. This was a carceral archipelago in which colonial rule differed from place to place but which remained quite distinct from metropolitan values and practices. By contrast, the repeal vision, produced by international and empire-wide humanitarian networks, authorized a geography of the British imperial system, in which colonial sites lost their specificity in their connections to each other and to the metropolis.

I would argue that the latter was, ultimately, the most powerful material and imaginative geography. The difficulties it faced cannot be downplayed – as Pope Hennessy's dilemma suggests – but in the end the humanitarian discourse triumphed. British rulers, in Hong Kong and elsewhere, were not able to continue regulating sex work by invoking local and cultural necessity. The decisive period, however, was the early 20th century, and the influence of the League of Nations. British colonial rulers – once again, not only in Hong Kong but elsewhere – had been put on the back foot by repeal pressure, and the repeal of the domestic Contagious Diseases Acts in 1886 led quickly to the suspension of the Hong Kong ordinance in 1888. The local administration in Hong Kong was convinced that this resulted in an immediate and alarming increase in venereal disease, but the reintroduction of regulationist legislation was quite impossible. In 1897 Colonial Secretary Joseph Chamberlain admitted to Governor Sir Henry Blake 'the fact that certain laws are or are not in force in the United Kingdom is not in itself a sufficient reason for applying or abolishing these laws in tropical Colonies, differing from the Mother Country in climate, race, social, moral, and religious conditions', but went straight on to add that reintroduction of regulation could not be sanctioned.[8] The colony was able to cook up a system by which the demands of propriety could be met, whilst keeping the substance of regulation: by promising to close brothels reported by the police, but making sure that the police only reported brothels that did not allow 'voluntary' medical inspections, a regulationist system remained in place into the 20th century.

Under the influence of League of Nations surveillance, however, it was impossible to continue in this regulationist vein. In 1931, colonial office officials admitted that, whilst technically illegal, brothels in the colony 'go on by consent, and, possibly, by implied condonation involved in our registration system'. There could, though, be no 'window-dressing' now.[9] The League of Nations, heavily influenced as it was by elements of abolitionism, feminism and social purity, had propagated a series of commissions on slavery, child welfare and the trafficking in women that all bore down heavily on systems of regulated prostitution. And although the same arguments were trotted out, by the same kinds of experts, these commissions succeeded in making regulationism internationally indefensible. The same 1931 minutes admitted as much, noting that

> Clearly, we ought to be very cautious assuming that western ideas on such matters are necessarily applicable without qualification to eastern communities, such as Hong Kong. On the other hand, we live in an age of League of Nations Commissions and are always in danger of being called to order if we fall short, in any part of the Empire, of Geneva standards.[10]

The League's methods – the use of confidential agents, for instance, on the ground, and of questionnaires distributed to metropolitan and colonial officials to ascertain the progress made in the campaigns against slavery and trafficking – echo those of the earlier repeal movements, but comprehensively transcended them in their ability to influence colonial governance.

The League's influence would only grow, during the inter-war years, and its policies would be taken up by the United Nations after the Second World War. The influence of international public opinion and the new global constitutional order is not well studied, and deserves far more scrutiny than it has so far received, but we can at least suggest here that studies of empire and colonial rule need to be properly complemented by an awareness of the international frameworks that existed even before the 20th century; it is wrong to assume that entities like the British Empire were either all-powerful or exclusive in this regard. The influence of international humanitarian networks needs to be considered within, beyond, and above imperial networks – so much so, by the early 20th century that we can talk I think of a 'meta-imperial' network in which British officials, in colonial sites and in the Whitehall, were enmeshed.

Conclusions

What I have tried to do in this chapter is to demonstrate the ways in which the colonial project of regulationism must be understood with regard to the geographies of the imperial network. On the one hand, I have shown how the regulation of prostitution in Hong Kong functioned within the wider framework of British imperial policies and preoccupations, producing a distinctive landscape of regulated sexuality. The nature and effects of the late 19th century contagious diseases ordinances have been explored with an eye to the local elaboration of colonial rule. The brothel districts of Victoria are distinctive and instructive colonial sites. On the other hand, however, I have cautioned against understanding the place of empire as localized or particular or indeed as *colonial*. I have contrasted the official discourse on regulationism, which insisted on regulationism as an essentially localised response to colonial conditions, with the humanitarian discourse of repealers who maintained that colonial sites and contagious diseases regimes were not only linked to one another but also to the metropolis. Their view, carried on into the 20th century, abjured the doctrine of local necessity, and put in its place a powerful view of an imperial system, even an imperial network. The distance they travelled may be measured by contrasting the view – outlined here by Stent's paper on the 'Chinese Social Evil' – of Hong Kong's regulationist regime as a borrowing from *Chinese* imperial custom, with that propagated by the National Council for Combating Venereal Diseases in their report on Hong Kong: 'We understand that the brothel is as much an offence to Chinese as to Western ethical teaching, and that concentrated promiscuity is a custom adopted from the West'[11] (see also MacPherson, 1997, pp. 97-99). It is this view, I have argued, that was ultimately dominant. Under the aegis of the 'meta-imperial network' of the League of Nations, the British system of regulated prostitution was exposed, condemned and dismantled. A hundred or so years after its initiation, the official registration and inspection of sex workers came to an end.

Notes

1   As Cooper (2004, p. 247) has commented in his review of recent work by Fergusson, Hardt and Negri, Kamen and Lieven, 'Empire is being discovered in the past, in the present, and in the future'.

2   This map includes contagious diseases legislation and related ordinances. British contagious disease legislation covered selected towns and garrisons in southern England and Ireland. The origins of Gibraltar's machinery for regulating prostitution are obscure.

3   In his review, Mark Harrison (2004, p. 483) claims, unconvincingly, that Levine considers 'the empire as an interconnected whole'. Levine (2003, p. 51) notes that contagious disease legislation was indeed 'empirewide', not merely a matter of local pragmatism, but would not endorse Harrison's claim.

4   Levine (2003, p. 6) rightly notes that 'British pride centred as much on a perceived distance from continental European mores as it did on separation from nonwhite peoples' but otherwise has no room for European colonies. Howell (forthcoming) discusses sexual regulation in nineteenth and early-twentieth century Gibraltar.

5   I have no problem here with the view of Ireland as a British *colony* – regulationism in Ireland has marked similarities with other colonial projects; for a good introduction to debates about Ireland's colonial status however see Fitzpatrick (1999), and also the more partisan revisionism of Howe (2000).

6   I am grateful here to Stephen Legg for sharing his understanding of the concept of governmentality.

7   In Ordinance 12 of 1857, registered brothels were confined 'within one of other of the following districts or portions of districts, namely, – Ha-wan, from Spring gardens,

eastward, – Sei-ping-poon, from the junction of Hollywood Road and Queen's Road West, westward, and Tai-ping-shan, except such parts of such districts or portions of districts as face the Queen's Road': see Hong Kong Acts, National Archives, Kew (NA), CO 130/2. Ordinance 10 of 1867 simply stipulated that 'the Registrar General may grant to any person to whom he shall think fit to keep a brothel in such district or other locality as the Governor in Council may from time to time appoint': see National Archives, Kew CO 130/3.

8 Chamberlain, J. (1897), memo to Governor Blake, 11 May 1897, National Archives, Kew, CO 882/6.
9 Colonial Office (1931), 'Position of Prostitution in Hong Kong', National Archives, Kew, CO 129/533/10.
10 Colonial Office (1931), 'Position of Prostitution in Hong Kong', National Archives, Kew, CO 129/533/10.
11 National Council (1921), *Report of the Commissioners of the National Council for Combating Venereal Diseases as to Conditions Affecting the Prevention and Cure of Venereal Diseases in Hong Kong, January, 1921*, National Archives, Kew, CO 129/472, pp. 317-413.

References

Adams, J.T. (1922), 'On the Term "British Empire"', *American Historical Review*, vol. 27, pp. 485-9.
Aisenberg, A. (2001), 'Syphilis and Prostitution: A Regulatory Couplet in Nineteenth-Century France', in R. Davidson and L.A. Hall (eds), *Sex, Sin and Suffering: Venereal Disease and European Society Since 1870*, Routledge, London, pp. 15-28.
Amos, S. (1877), *A Comparative Survey of Laws in Force for the Prohibition, Regulation, and Licensing of Vice in England and Other Countries*, Stevens and Sons, London.
Andrew, E.W. and Bushnell, K.C. (1898), *The Queen's Daughters in India*, Morgan & Scott, London.
Ballhatchet, K. (1980), *Race, Sex and Class Under the Raj: Imperial Attitudes and Policies and Their Critics, 1793-1905*, Weidenfeld and Nicolson, London.
Banerjee, S. (1998), *Under the Raj: Prostitution in Colonial Bengal*, Monthly Review Press, New York.
Baucom, I. (1999), *Out of Place: Englishness, Empire and the Locations of Identity*, Princeton University Press, Princeton.
Benyon, J. (1991), 'Overlords of Empire? British "Proconsular Imperialism" in Comparative Perspective', *Journal of Imperial and Commonwealth History*, vol. 19, pp. 164-202.
Brown, D.B. (1888), 'The Pros and Cons of the Contagious Diseases Acts, as Applied in India', *Transactions of the Medical and Physical Society of Bombay*, vol. 11, pp. 80-97.
Burroughs, P. (1999), 'Imperial Institutions and the Government of Empire', in A. Porter (ed.), *The Oxford History of the British Empire, Volume III, The Nineteenth Century*, Oxford University Press, Oxford, pp. 170-97.
Bryder, L. (1998), 'Sex, Race and Colonialism: An Historiographical Review, *International History Review*, vol. 20, pp. 791-822.
Canny, N. (1998), 'The Origins of Empire: An Introduction', in N. Canny (ed.), *The Oxford History of the British Empire, Volume 1, The Origins of Empire: British Overseas Enterprise to the Close of the Seventeenth Century*, Oxford University Press, Oxford, pp. 1-33.

Chamberlain, J. (1897), Memo to Governor Blake, 11 May 1897, National Archives, Kew, CO 882/6.

Colonial Office (1931), 'Position of Prostitution in Hong Kong', National Archives, Kew, CO 129/533/10.

Cooper, F. (2004), 'Empire Multiplied', *Comparative Studies in Society and History*, vol. 46, pp. 247-72.

Cooper, F. and Stoler, A.L. (1997), 'Between Metropole and Colony: Rethinking a Research Agenda', in F. Cooper and A.L. Stoler (eds), *Tensions of Empire: Colonial Cultures in a Bourgeois World*, University of California Press, Berkeley, pp. 1-56.

Corbin, A. (1990), *Women For Hire: Prostitution and Sexuality in France After 1850*, Harvard University Press, Cambridge, MA.

Deakin, C.W.S (1872), *The Contagious Diseases Acts: The Contagious Acts '64, '66, '68 (Ireland), '69. From a Sanitary and Economic Point of View*, H.K. Lewis, London.

Firth, C.H. (1918), '"The British Empire"', *Scottish Historical Review*, vol. 15, pp. 185-9.

Fitzpatrick, D. (1999), 'Ireland and the Empire', in A. Porter (ed.), *The Oxford History of the British Empire, Volume III, The Nineteenth Century*, Oxford University Press, Oxford, pp. 495-521.

Gikandi, S. (1996), *Maps of Englishness: Writing Identity in the Culture of Colonialism*, Columbia University Press, New York.

Grewal, I. (1996), *Home and Harem: Nation, Gender, Empire, and the Cultures of Travel*, Duke University Press, Durham, NC.

Hall, C. (1994) 'Rethinking Imperial Histories: The Reform Act of 1867', *New Left Review*, vol. 208, pp. 3-29.

Hall, C. (2002) *Civilising Subjects: Metropole and Colony in the English Imagination, 1830-1867*, Polity, Oxford.

Hammond, J. and Hammond, B. (1932), *James Stansfeld: A Victorian Champion of Sex Equality*, Longmans, London.

Harrison, M. (2004), Review of Philippa Levine, *Prostitution, Race, and Politics*, *American Historical Review*, vol. 109, pp. 483-4.

Harsin, J. (1995), *Policing Prostitution in Nineteenth-Century Paris*, Princeton University Press, Princeton.

Hong Kong (1879), *Report of the Commissioners Appointed By His Excellency John Pope Hennessy, C.M.G., Governor and Commander-in-Chief of the Colony of Hong Kong and its Dependencies, &c., &c., &c., to Enquire into the Working of "The Contagious Diseases Ordinance, 1867"*, Noronha, Hong Kong.

Hong Kong (1882), *Correspondence Respecting the Alleged Existence of Chinese Slavery in Hong Kong*, Eyre and Spottiswoode, London.

Howe, S. (2000), *Ireland and Empire: Colonial Legacies in Irish History and Culture*, Oxford University Press, Oxford.

Howell, P. (2000), 'Prostitution and Racialised Sexuality: The Regulation of Prostitution in Britain and the British Empire before the Contagious Diseases Acts', *Environment and Planning D: Society and Space*, vol. 18, pp. 321-39.

Howell, P. (2003), 'Venereal Disease and the Politics of Prostitution in the Irish Free State', *Irish Historical Studies*, vol. 33, pp. 320-41.

Howell, P. (2004), 'Sexuality, Sovereignty and Space: Law, Government and the Geography of Prostitution in Colonial Gibraltar', *Social History*, vol. 29, pp. 444-64.

Jacobs, J. (1996), *Edge of Empire: Postcolonialism and the City*, Routledge, London.

JanMohamed, A. (1992), 'Sexuality on/of the Racial Border: Foucault, Wright, and the Articulation of "Racialized Sexuality"', in D. Stanton (ed.), Discourses of Sexuality: From Aristotle to AIDS, University of Michigan Press, Ann Arbor, pp. 94-116.

Jaschok, M. and Miers, S. (1994), 'Women in the Chinese Patriarchal System: Submission, Servitude, Escape and Collusion,' in M. Jaschok and S. Miers (eds), *Women and Chinese Patriarchy: Submission, Servitude and Escape*, Zed, London, pp. 1-24.

Kalpagan, U. (2000), 'Colonial Governmentality and the "Economy"', *Economy and Society*, vol. 29, pp. 418-38.

Kalpagan, U. (2002), 'Colonial Governmentality and the Public Sphere in India', *Journal of Historical Sociology*, vol. 15, pp. 35-58.

Lambert, D. and Howell P. (2003), 'John Pope Hennessy and the Translation of "Slavery" between Late Nineteenth-Century Barbados and Hong Kong', *History Workshop Journal*, vol. 55, pp. 1-24.

Lester, A. (2001), *Imperial Networks: Creating Identities in Nineteenth Century South Africa and Britain*, Routledge, London.

Lester, A. (2002), 'Obtaining the "Due Observance of Justice": The Geographies of Colonial Humanitarianism', *Environment and Planning D: Society & Space*, vol. 20, pp. 277-93.

Lester, A. (2002a), 'British Settler Discourse and the Circuits of Empire', *History Workshop Journal*, vol. 54, 27-50.

Levine, P. (1994), 'Venereal disease, Prostitution, and the Politics of Empire: The Case of British India', *Journal of the History of Sexuality*, vol. 4, pp. 579-602.

Levine, P. (1998), 'Modernity, Medicine, and Colonialism: The Contagious Diseases Ordinances in Hong Kong and the Straits Settlements', *Positions*, vol. 6, pp. 675-705.

Levine, P. (2000), 'Orientalist Sociology and the Creation of Colonial Sexualities', *Feminist Review*, vol. 65, pp. 5-21.

Levine, P. (2003), *Prostitution, Race and Politics: Policing Venereal Disease in the British Empire*, Routledge, London.

Louis, W.M. (Ed, 1998-), *The Oxford History of the British Empire*, 5 volumes, Oxford University Press, Oxford.

Lowe, K. and McLaughlin, E. (1992), 'Sir John Pope Hennessy and the "Native Race Craze": Colonial Government in Hong Kong 1877-1882', *Journal of Imperial and Commonwealth History*, vol. 20, pp. 223-47.

Luddy, M. (1997), '"Abandoned Women and Bad Characters": Prostitution in Nineteenth Century Ireland', *Women's History Review*, vol. 6, pp. 485-503.

McLintock, A. (1995), *Imperial Leather: Race, Gender and Sexuality in the Colonial Contest*, Routledge, London.

MacPherson, K.L. (1995), 'Caveat Emptor! Attempts to Control the Venereals in Nineteenth Century Hong Kong,' in L. Bryder and D.A. Dow (eds), *New Countries and Old Medicine*, Auckland Medical History Society, Auckland, pp. 72-78.

MacPherson, K. (1997), 'Conspiracy of Silence: A History of Sexually Transmitted Diseases and HIV/AIDS in Hong Kong', in M. Lewis, S. Bamber and M. Waugh (eds), *Sex, Disease, and Society: A Comparative History of Sexually Transmitted Diseases and HIV/AIDS in Asia and the Pacific*, Greenwood Press, Westport, pp. 85-112.

Manderson, L. (1997), 'Colonial Desires: Sexuality, Race, and Gender in British Malaya', *Journal of the History of Sexuality*, vol. 7, pp. 372-88.

Miners, N. (1987), *Hong Kong Under Imperial Rule 1912-1941*, Oxford University Press, Hong Kong.

Munn, C. (2001), *Anglo-China: Chinese People and British Rule in Hong Kong, 1841-1880*, Curzon, Richmond.

National Council (1921), *Report of the Commissioners of the National Council for Combating Venereal Diseases as to Conditions Affecting the Prevention and Cure of Venereal Diseases in Hong Kong, January, 1921*, National Archives, Kew, CO 129/472, pp. 317-413.

Ogborn, M. (1993), 'Law and Discipline in Nineteenth Century English State Formation: The Contagious Diseases Acts of 1864, 1866, and 1869', *Journal of Historical Sociology*, vol. 6, pp. 28-55.

Parliamentary Papers GB (1871), *Report from the Royal Commission on the Administration and Operation of the Contagious Diseases Acts 1866-69*, PP 1871 (C.408) XIX.

Pedersen, S. (2001), 'The Maternalist Moment in British Colonial Policy: The Controversy over 'Child Slavery' in Hong Kong, 1917-1941', *Past and Present*, vol. 171, pp. 161-202.

Scott, D. (1995), 'Colonial Governmentality', *Social Text*, vol. 43, pp. 191-220.

Sinha, M. (2002), '"Signs taken for wonders": the stakes for imperial history', *Journal of Colonialism and Colonial History*, vol. 3, http://www.press.jhu.edu/journals/journal_of_colonialism_and_colonial_history/.

Stansfeld, J. (1882), *Lord Kimberley's Defence of the Government Brothel System at Hong Kong*, National Association for the Repeal of the Contagious Diseases Acts, London.

Stoler, A.L. (1996), *Race and the Education of Desire: Foucault's History of Sexuality and the Colonial Order of Things*, Duke University Press, Durham.

Stoler, A.L. (2002), *Carnal Knowledge and Imperial Power: Race and the Intimate in Colonial Rule*, University of California Press, Berkeley.

Taithe, B. (1992), From Danger to Scandal: Debating Sexuality in Victorian England: The Contagious Diseases Acts (1864-1869) and the Morbid Imagery of Victorian Britain, unpublished PhD, Department of History, University of Manchester.

Van Heyningen, E.B. (1984), 'The Social Evil in the Cape Colony 1868-1902: Prostitution and the Contagious Diseases Acts', *Journal of Southern African Studies*, vol. 10, pp. 170-91.

Walkowitz, J.R. (1980), *Prostitution and Victorian Society: Women, Class and the State*, Cambridge University Press, Cambridge.

Ware, H. (1969), Prostitution and the State: The Recruitment, Regulation, and Role of Prostitution in the Nineteenth and Twentieth Century, unpublished PhD, Department of Sociology, University of London.

Warren, J.F. (1993), *Ah Ku and Karayuki-San: Prostitution in Singapore 1870-1940*, Oxford University Press, Singapore.

Whitehead, J. (1995), 'Bodies Clean and Unclean: Prostitution, Sanitary Legislation, and Respectable Femininity in Colonial North India', *Gender & History*, vol. 7, pp. 41-63.

Wills, C. (1996), 'Joyce, Prostitution, and the Colonial City', *South Atlantic Quarterly*, vol. 95, pp. 79-95.

Wilson, H.J. (1907), Copy of a Rough Record of Events and Incidents connected with the Repeal of the "Contagious Diseases Acts, 1864-69," in the United Kingdom, and of the Movement against State Regulation of Vice in India and the Colonies, 1858-1906, Parker Bros., Sheffield.

PART III
'DISPLACEMENT'

Chapter 11

Displacement

Lindsay Proudfoot and Michael Roche

The diverse colonial places and imperial spaces explored in the contributions to this volume aptly reflect the contingent and heterogeneous character of the colonialist experience of Empire. More immediately, they remind us of the impossibility of providing a straightforward answer to the question that was posed in chapter one: 'when, where, and what was Empire?' The subjectivities of the men and women of predominantly European descent, whose variously empowered lives have been recovered here, alert us to the inchoate and fractured cultural meanings inscribed in place and space by settler and other colonialist communities within Empire.[1] If as Ania Loomba notes (1998, p. 179), 'the colonialist presence was felt very differently by various subjects of Empire', then that presence was also constructed very differently among the colonialists themselves. Their experience of Empire was contingent on their particular constitutive encounter with the existing structures, agencies and practices of colonialism, which were themselves riven through with discourses of gender, class and ethnicity. These destabilized the 'straightforward' racialized binary of the 'white' presence in Empire, which was in any case variably inscribed, and exposed the creative spaces through which settlers and other colonialists might negotiate similarly diverse and ambiguous spaces of meaning with indigenous collectivities (Rose and Blunt, 1994; Stasiulis and Yural-Davis, 1995). Thus materially-bounded sites of colonization were continually 'displaced' in their meaning by these subversive renderings of multiple and fractured identities.

In emphasizing the spatiality and materiality, as well as the diversity of settler and other colonialist experience, the chapters in this collection attempt to counter the under-representation of these aspects of Empire by postcolonial scholarship in which accounts of colonialist experience have been inflected by an ahistorical synecdochy (Johnson, 2003). Arguably, this has 'flattened' representations of the colonialist encounter with Empire, just as it has helped to erase the nuanced complexities of space and place which were constitutive of this. A note of caution is in order here, however. One of the fundamental geographical fractures within that *congerie* of dominions, protectorates and colonies which is subsumed under the title of the 'British Empire', and which the chapters in this collection attempt to bridge, is that between the so-called colonies of settlement, the 'white dominions' of Australia, Canada, New Zealand and South Africa, and the colonies of Empire or exploitation. India provides a prime example of the latter. Here, the colonialist presence was arguably less pervasive, or at least, intrusive in different ways, compared to the specifically settler presence in the 'white dominions'. Thus, any apparent postcolonial insensitivity towards the heterogeneity of colonialist and settler materiality and spatiality may be accounted for by its primary concern with the colonies of exploitation. In these parts of Empire, indigeneity maintained a cultural presence in more powerful and potent ways than in most of the 'white dominions' (New Zealand being one exception). Nevertheless, in their focus on Ireland and Australasia as well as India and Hong Kong, the contributions to this book recover a rich diversity of colonialist experience, and this formed a fundamental, but far from necessarily disabling, part of the fabric of Empire for people of all races, colours and creeds. Thus in order to further our understanding of colonialism and its consequences, we must engage with the subjectively constructed material worlds of the coloni*zer* just as much as with those of the coloni*zed*, while recognizing that the former will have been differentially constructed depending on the relative presence or absence of indigeneity.

Having acknowledged the importance of these indigenous contexts in the overall construction of colonialist identities, we return to the specific issue of how different settler and other colonialist identities were expressed in spatial and material terms, particularly in those parts of Empire where racialized indigeneity occupied less of a foreground position than in the colonies of exploitation. At the heart of this issue lies the relationship between the polyvocality of place and the hybridization of cultural identities. It is in this interaction that we find the 'displacement' alluded to at the start of this chapter. As socially constructed and materially bounded sites of 'memory, agency and identity', colonial places supported social and cultural meanings that were both intensely personal but also shared by other groups or individuals, similarly positioned within the prevailing imperial/colonial formation. But equally, they might also support contrary meanings invested in them by those use of these same shared spaces reflected their different positionality within the colonial formation. In this way, places were polyvocal. That is, they supported a variety of symbolic meanings constructed through the agency of those people whose life paths intersected through them, but which were interpreted in ways that depended on the positionality of the 'reader'.

It is, of course, easy to envisage how this sort of 'oppositional' polyvocality might have given voice to racialized hegemonic and subaltern groups within the colonies of exploitation. As Howell and Kumar have shown, the spaces of regulated prostitution in 19th century Hong Kong and India were read variously as sites of legalized immorality or social necessity by reforming and regulatory groups within the hegemonic colonial class, but also as sites of racially transgressive caste (mis)behaviour by indigenous Indians. But it is important to realize that, as Bhabha (1994) has argued, these seemingly discordant voices in fact articulated identities that were mutually constructed. Whelan's account of Queen Victoria's visit to Dublin in 1900 provides a case in point: the fluid spaces of imperial/colonial celebration in the city were also spaces of resistance. Accordingly, each 'site' of celebration/resistance was read in terms of a very different, but interdependent, symbolism: loyalty and empowerment on the one hand, oppression and marginalization on the other. Each required the other as its referent. What was loyalty if there was no disloyalty? What were nationalist aspirations if there was no colonial rule? In the same way, as Duffy shows, Ireland's landed estates also embodied deeply contested political, cultural and ethnic meanings that were so intermeshed it is difficult to disentangle them. Thus, while the country's landlords might be expected to have shared a general understanding of the social and other vocabularies embodied in their peers' demesnes and 'big houses' which, being derived from their own positionality was uniquely their own, it was precisely the *fact* of this discourse of elite 'Otherness' that was the focus for nationalist agrarian resistance. As subjective sites of meaning, these estates gave a common materiality to mutually antagonistic ideologies that were, in fact, mutually dependent.

In both these instances, the consequence of this constructive mutuality of meaning was the creation of a form of hybridized symbolic space. Narratives of landownership in 19th century Ireland could only be inscribed in place in the context of underlying, subversive, agrarian hostility and resistance; the politicized spaces of domestic Irish nationalist aspiration could only be constructed within the defining materiality of the colonial/imperial state. The other 'subjective material worlds' explored in this collection of chapters support similar interpretations of mutual construction, suggesting that certain forms of cultural hybridity were as fundamental to the spatial inscription of specifically settler identities as they were to material colonial practice in Ireland or, the focus of Bhabha's concern, the racialized transcultural contact zone between the colonizer and colonized in colonies of exploitation. Bhabha emphasizes the interdependence of the colonizer and colonized, and the manner in which their 'mutually constructed subjectivities' create an ambiguous relational 'third-space', characterized by 'ambivalence' and 'mimicry', within which all colonial exchanges take place (Ashcroft *et al.*, 1998; McLeod, 2000). The notion of ambivalence is a particularly powerful one in the context of racialized transcultural contacts, but also finds echoes in Proudfoot's account of Presbyterian discourse in colonial Australia. The term refers to the 'complex mix of repulsion and attraction that characterizes the relationship between the colonized and the colonizer'. Never completely resistant or complicit within the colonization process, the colonized subject fluctuates between these positions, becoming at most an imperfect simulacrum of the colonizer. Seemingly compliant, the colonized subject's 'westernized' demeanour contains within itself the seeds of colonialism's own subversion. Too perfect an approximation to Western *mores* by the colonial subject, and the colonial project itself becomes challenged; but equally, imperfection leads to parody and mimicry, which in the end is equally destabilizing (Ashcroft *et al.*, 1998).

If for Western colonization in general, we substitute the process of colonial Presbyterian church foundation, then similar ambivalences may be detected in the spaces of meaning erected in its wake. This ambivalence operated at various levels. In the example presented here, it is evidenced in the personal understandings of place which underpinned William Hamilton's sense of colonial selfhood, and in the more broadly drawn relationship between the colonial Presbyterian Church and the Established Church of Scotland. Thus a significant proportion of the Presbyterian congregation at Goulburn was initially drawn to, but subsequently repelled by, Hamilton's particularly austere brand of Presbyterian theology; while the New

South Wales Synod in general was equally divided in its response to the Disruption and the conflicting claims of the Free and Established Scottish Churches on its allegiance. The result in each case was a new form of institutional hybridity. At Goulburn, a divided congregation redefined its denominational allegiance, but only after it been distanced spiritually by the minister its wealthier members had rejected. In Sydney, the governing Church body also redefined its allegiances and its place within the discursive religious networks of Empire, but only at the cost of exacerbating its underlying (and continuing) internal theological divisions.

Hybridity of a different sort characterized the gendered spaces of Glenorchy in the 1860s, Feminist studies of Empire have emphasized the ambiguous position of white women in Bhabha's 'third space of encounter'. Although still marginalized within the discourse of colonial authority by their traditionally assigned gender status, many were simultaneously newly empowered in their relationship with the indigenous 'Other' by their racial identity. In this way, the circuits of their authority might move beyond traditionally 'feminized' domestic space (Rose and Blunt, 1994). The example of Glenorchy demonstrates that similar boundaries to gendered behaviour existed (to be transgressed) in the 'white dominions' that were not contingent on an indigenous presence. As Hall's example demonstrates, the transgressive behaviour might equally be a matter of social status, and the boundaries those prescribed by normative expectations of social behaviour within the local settler community.

This concept of transgressive identities constructed through the disruption and reinscription of a signifying memory of previously bounded experience, lies at the heart of postcolonial understandings of cultural hybridity. Closely associated with it is the idea of 'diaspora', conceived of not as a necessarily forced exile – its original meaning – but as a multi-generational community of common migrant origin, for whom some shared sense of that origin still operates to structure their sense of selfhood. Crucially, however, this sense of their shared 'roots' is cross cut and modified by their ever changing, and individually different, experience of the 'routes' through which their present circumstances came to be constructed. In this way, diasporic communities are increasingly estranged from their original identity. In their ever changing diasporic 'present', the 'boundaries' inscribed to their identity by their cultural rootedness are continuously disrupted by the 'introduction of often incommensurable cultural temporalities, …(their shared) uncontrollable (and more recent diasporic) pasts that inhabit (their) present' (Bhabha, 1994; McLeod, 2000).

This sense of diasporic identity is particularly privileged in three of the chapters in the collection: McCann's account of settler engagements with the landscapes of the New South Wales Wheat belt; Powell's discussion of environmental understandings and the formation of a 'national' identity in Australia; and Roche's analysis of the nationalist/imperialist content of the Soldier Settlement debate in New Zealand during and after the First World War. Although offering perspectives on very different time/place moments of Empire, all three authors demonstrate the centrality of 'routed' rather than 'rooted' identity constructions to their narratives. Thus the settler-descended farmers of the Lachlan Valley continue to renegotiate their identity – their sense of who and what they are – in terms of their perceived need to come to a new understanding of the landscape they have helped shape – and their forebears appropriated – in light of earlier European (mis)understandings of what the land could sustain. Perhaps ironically, but in any case positively, these new engagements offer a belated recognition of aboriginality's earlier encounters with the same landscapes. For both Powell and Roche, the diasporic 'routes' of identity created new understandings of what it was to be Australian or a New Zealander. Powell argues that in colonial and Federation Australia, this occurred through a process of environmental learning and a coming to terms with the unique 'Otherness' of the continent's physical heritage, that demanded a recalibration of land-based 'national' narratives. In New Zealand, Roche's focus on the Soldier settlement debate demonstrates that here, too, the narrative of national identity was diasporic, and by the early 20th century, involved a morally-charged renegotiation of the country's imperial heritage in the light of a re-imagining of its national future.

In their various ways, therefore, the chapters brought together in this book offer the beginnings of one possible perspective on the historical geographies of settler and other colonialist experience of Empire. By emphasizing the diverse localism of this experience, yet taking care to place this within the context of the broader discursive flows of people, ideas and material which articulated Empire as cultural practice, the contributors have demonstrated both the complexity of these geographies and, by implication, the inadequacy of totalising 'meta-narratives' which seek to explain them. The geographies the contributors have recovered are, for the most part, explicitly and intentionally those of some of the 'white' subjects of Empire. These geographies were heterogeneous in their material construction and ambiguous in their meaning. Characterized by hybridity rather than essentialism, their meanings and the cultural identities these inscribed were variously expressed: in diasporic terms; in the confused and ambiguous interactions with racially constructed indigeneity; or, in the case of Ireland, in the intimate ethnic spaces created by that country's part complicity and part subordination within British imperialism. This diversity reminds us that if Empire was a globalizing phenomenon, it was also a local one.

Note

1. 'Settlers' are defined as those who emigrated with the intention of remaining permanently in their eventual country of destination. 'Colonialists' are conceived of as including these, but as also referring to the broader group of shorter-term residents such as soldiers, administrators, missionaries, and the colonial police and judiciary.

References

Ashcroft, B., Griffiths, G. and Tiffin, H. (1998), *Key Concepts in Post-Colonial Studies*, Routledge, London.
Bhabha, H.K. (1994), *The Location of Culture*, Routledge, London.

Blunt, A. and Rose, G. (1994), *Writing Women and Space. Colonial and Postcolonial Geographies*, The Guilford Press, New York.

Johnson, R. (2003), *British Imperialism*, Palgrave Macmillan, Basingstoke.

Loomba, A. (1998), *Colonialism/Postcolonialism*, Routledge, London.

Mcleod, J. (2000), *Beginning Postcolonialism*, Manchester University Press, Manchester.

Stasiulis, D. and Yuval-Davis, N. (eds) (1995), *Unsettling Settler Societies. Articulations of Gender, Race, Ethnicity and Class*, Sage Publications, London.

Index